ENVIRONMENTAL
JUSTICE
IN AMERICA

Environmental Justice in America

A New Paradigm

Edwardo Lao Rhodes

INDIANA
University Press
Bloomington & Indianapolis

This book is a publication of
Indiana University Press
601 North Morton Street
Bloomington, IN 47404-3797 USA

http://iupress.indiana.edu

Telephone orders   800-842-6796
Fax orders   812-855-7931
Orders by e-mail   iuporder@indiana.edu

© 2003 by Edwardo Lao Rhodes

The paper used in this publication meets the minimum requirements of American National Standard for Information Sciences—Permanence of Paper for Printed Library Materials, ANSI Z39.48-1984.

Manufactured in the United States of America

Library of Congress Cataloging-in-Publication Data
Rhodes, Edwardo Lao, date
Environmental justice in America : a new paradigm / Edwardo Lao Rhodes.
p. cm.
Includes bibliographical references and index.
ISBN 0-253-34137-X (cloth : alk. paper)
1. Environmental justice—United States. I. Title.
GE230 .R48 2003
363.7′00973—dc21
2002006589

1 2 3 4 5   08 07 06 05 04 03

Contents

Preface and Acknowledgments   vii

PART I

The Dynamics of Environmental Justice

One: Introduction   5

Two: Forms of Environmental Justice   13

Three: What Has Gone Before: Why Race Was
Not on the Original Environmental Agenda   30

Four: The Evolution of Environmental Justice
as a Policy Issue: A Movement Whose Time
Has Come   43

Five: Misconceptions about Minority Attitudes
toward Environmental Issues   72

Six: The EPA: An Agency with an Attitude   86

PART II

Policy Analysis of Environmental Justice

Seven: Environmental Justice through the Lens
of Policy Analysis: Why Should Government
Get Involved?   105

Eight: The Measurement of Environmental
Justice: Some Rules of Engagement   118

Nine: A New Way of Looking at the Same Old
Numbers: Using Data Envelopment Analysis to
Evaluate Environmental Quality   137

Contents

vi

PART III

A Case, a Summary, and Some Conclusions

Ten: A Case of Environmental Justice: The
Disposal of Hazardous Material in Noxubee
County, Mississippi   161

Eleven: Policy Directions and Recommendations   186

Twelve: Environmental Justice: A New Paradigm
—A Time of Change   206

Appendix: Principles of Environmental Justice   213

Notes   217

Bibliography   245

Index   257

Preface and Acknowledgments

As a public-sector policy analyst with a background in both operations research and economics, I have had a special interest in environmental and natural resource policy. In pursuing that interest, I have always felt a need to bring social and economic policy sensitivity to a field that historically has been woefully insensitive to such dynamics. Nevertheless, in spite of such intentions, I had never even heard the terms *environmental racism, environmental equity,* or *environmental justice* until just a few years ago. Until very recently, the environmental policy area simply had not made a place for this topic on its agenda of issues. Despite a background that should have sensitized me to such issues, I had difficulty breaking out of the field's conceptual constraints. I did not recognize until quite recently that an entire dimension of environmental policy, regulation, and activity has been utterly ignored.

Once I began reading the small but growing literature that existed on the subject of unequal environmental impacts, I felt ashamed that it had taken me so long to consider the obvious social justice dynamics and the implicit policy significance of the matter. It is a tenet of public-policy studies that virtually any public or private activity—from education financing to the information superhighway, from public television to corporate downsizing—will have very different impacts on different social, economic, and racial categories. Yet amazingly, in environmental policy literature and practices, such differentiation has historically not been recognized.

At the same time, during the last twenty-five years, public and private environmental activities and policies have assumed an ever more important role in our lives, affecting everything from what we breathe and where, to what we build and how. Few would deny that the effects, both positive and negative, of these activities and policies are unequally distributed across various segments of society. This inequality even suggests that some segments of our society may have benefited at the expense of others.

While accepting the likelihood of distributional inequity, many people, both inside and outside the environmental movement, have been either unwilling or unable to take the next intellectual step: the admission that the consideration of the unequal consequences of such environmental decision-making merits at least as much concern as environmental issues like the level of pollutants in our drinking water or the threat of global warming.

Failure to recognize environmental inequity has extended even into our academic and intellectual centers. For example, I have taught for the last seventeen years at a major research and teaching university in a school with a set of professional degree programs, one of whose primary missions is environmental policy study and education. Yet until 1992, decades after the school's founding, issues of environmental justice simply were not explored. From an inspection of our curriculum offerings, one would get the impression that race, class, and ethnicity were of no consequence in environmental science or policy. I offer these comments not as a condemnation of my home institution but as an observation about the entire field. With few exceptions, at the beginning of the 1990s, one would have had to look long and hard to find any treatment or even acknowledgment of the subject of environmental justice in the curricula of the major schools in this country that specialize in natural resource management, environmental science, environmental policy, or public policy and affairs of a country.[1]

This lack of attention did not come, I believe, from some innate insensitivity to the conditions of the poor or of racial minorities. My school, like many other public-policy institutions during the same decades, devoted considerable time and resources to examining social policy issues in other public-sector areas such as education, criminal justice, and health care. Nevertheless, in the area of environmental policy, this connection simply had not been made—by us or by many others in the profession. I suspect that intellectual compartmentalization allowed me, and perhaps many of my colleagues, to worry separately about environmental policy issues, and about social and economic injustice, but we failed to see or to act on the obvious connections between the two.

As I studied the existing literature and writings on environmental justice, I soon realized how little was known about environmental inequity. Almost since the beginning of the modern environmental movement in the late 1960s, there has been some research on race-based differences in environmental risk, but the number of such works was small, and the topic treated was of no more than peripheral importance in the broader spectrum of environmental policy concerns. Except for a few heroic pioneering works by groups such as the United Church of Christ[2] and a few isolated scholars such as Robert Bullard,[3] very little empirical work directly investigated this issue before the late 1980s.[4] And when the subject was studied or investigated, it was considered most often as a public health problem rather than an environmental problem.[5]

When I began this book, my primary intention was to write a primarily quantitative treatment of environmental justice demonstrating how the powerful tools of operations research and economics could be brought to bear on various aspects of the issue. The environmental-justice question is a gigantic challenge to public-policy analysis, particularly quantitative policy analysis. From the need to combine environmental risk and location information with area-specific demographic information, to trying to assess cumulative risk at the individual neighborhood level, this area will provide fresh quantitative challenges for quite some time. In fact, in many ways, the formulation and implementation of environmental-justice policy directly depends on the rigorous application of powerful quantitative analyses.

During a year's sabbatical, which I spent at the U.S. Environmental Protection Agency's (EPA) newly created Office of Environmental Justice, I intended to collect as much data and run as many analyses as I could on selected dimensions of environmental justice. I had not intended to wander into the more treacherous territory of issue development. Whether by fortunate opportunity or natural progress, my focus changed, and I could not ignore the chance to explore the broader dynamics of this topic. The field of environmental justice is still at such an early stage of development that this more fundamental issue demanded at least as much of my attention as my original, more quantitative, interests. This work is thus a synthesis of the need to explore general issues and a careful application of powerful quantitative tools and constructs from public-policy analysis.

In trying to shine light on the broader questions surrounding the environmental-justice question, I was struck by how that question fits into the more general issue of social and economic impacts of environmental decisions. In many ways, the neglect of the specific issue of environmental justice reflects the general neglect or avoidance of social and economic impact issues within the field of environmental policy. This neglect, as will become clearer later in this work, is a direct outcome of the roots of the modern environmental movement.

We embark on a journey to discover the nature and extent of environmental injustice. However, we will also explore the broader, and perhaps in the end more fundamental, issue of the social policy of differentiated environmental impacts. As with most explorations, this one will also uncover many new questions that will require yet further investigation.

In this book, I challenge those in the environmental movement to consider how and why such a major aspect of environmental policy and activity has been ignored for so long. Perhaps as significant, I also explore what the emergence of environmental justice as a major public-policy issue means for the environmental field as a whole. For the public-policy analyst, the book will introduce a whole new area of policy issues that begs for further study. At the same time, my examination of practical approaches to investigating claims of environmental inequity, organizing guides, and methodologies will provide one approach on which to base that future exploration. Finally, for the policy-maker, especially at the local or state level, the specification of broad guidelines for environmental-justice policy offers a way to shape future decisions or modify past decisions.

At the start of the present journey, however, I offer a few words of guidance. First, I advise readers with strong environmental ties to jettison the extra weight of defending the entire environmental movement against charges of environmental racism or social insensitivity. The journey will be more comfortable, less trying, if each step is taken by itself with as few preconceptions as possible.

For those readers with strong connections to environmental justice or civil rights, the same advice applies. This exploration will be more useful and beneficial if each step is taken without presupposing that a culprit has to be identified with every negative or skewed environmental outcome. All past decisions and activities, however negative their impact on some segments of our society, need not be the result of a conspiracy to exploit.

Finally, for readers who believe that market forces explain everything: Resist the urge to explain away every instance of environmental inequity as an unavoidable consequence of forces beyond the control of environmental policy- and decision-makers. Remember that ultimately markets are people, not numbers or abstract dynamics, and in most cases, people should be held responsible for their actions. Market failures do occur, and environmental inequity and injustice can be best considered, regardless of one's starting position, with as open a mind as possible. The subject has a deep emotional dimension, and people on all sides, many who purport to know better, have tended to respond with cries of indignation rather than with studied and focused words. I am aware that environmental justice as an important social issue was not conceived in the halls of ivy, nor in the corridors of a government agency. I further admit that virtually no major social experiment or policy issue has ever been started with a university study or treatise. Yet, while not denying the importance of those cries of indignation, I maintain that viable environmental-justice solutions and remedies require studied and focused words.

Two institutions in Washington, D.C., extended me great opportunities to explore this topic. First, my thanks go to the EPA's Office of Environmental Justice; its former director, Clarice Gaylord, and her staff made me feel both welcomed and part of a truly dedicated team. My thanks also go to the School of Public Policy at George Mason University, and especially its dean, Kingsley Haynes, a longtime friend and colleague. The school provided a temporary academic home where I could write and explore in peace, and I welcomed the much-needed assistance that a research university provides so well.

Finally, three people must be singled out for special commendation. First, Robert Sloan, Indiana University Press senior sponsoring editor, has remained steadfast in his faith in my book. His calm demeanor and quiet confidence pulled me through more than one bout of self-doubt. Second, Bob Furnish, my copy editor, has suffered through at least three, and in some places four, revisions of the book. Without much complaint, Bob worked to correct my most egregious writing errors, and more important, he provided a needed everyman perspective in his reading and interpretation of the book. Third, Kim Shipley, my secretary, has been loyal, timely, and, most important, extremely efficient in getting the book into its current form. Without the assistance of these special people, the book may have never been more than a good idea. To all three of them, sincere and heartfelt thanks.

Part I

The Dynamics of Environmental Justice

In the chapters of Part 1, we explore the key dimensions of the environmental-justice issues. Given their early stages of development, a necessary exercise is establishing a working set of terms for discussion. After the Chapter 1 introduction of the topic, Chapter 2 provides short expositions and illustrations defining key terms in this debate, such as *environmental racism, environmental equity, environmental justice,* and *environmental protection rights.* As Chapter 2 explains, much misunderstanding and misinterpretation of conditions in this area directly result from participants hearing the same terms but assigning them different meanings and significance. While the definitions provided are neither definitive nor exhaustive, they do establish a fixed starting point for subsequent discussion. A very important feature of Chapter 2 will be specifying a matrix classification of environmental-justice issues based on whether the issues are geographically and location specific or population and nonlocation specific. Confusion over "environmental-justice remedies," or even how to analyze a particular environmental-justice problem, often springs from failing to appreciate the inherent difference between types of environmental-justice issues.

Chapter 3 explores the question, Why were race and other factors of societal stratification left out of the original environmental agenda? Our initial focus in this chapter will be on the early preservation/conservation roots of the modern environmental movement and what those roots mean in terms of modern attitudes and agendas. A case is made that a major reason for the modern agendas' neglect can be traced backed to the founda-

1

tion philosophy and early beginnings of the modern environmental move-ment. It is argued that many of the questioned behavior and policies en-countered in environmental organizations have less to do with some de-liberate racism or social elitism in the movement and more to do with these organizations having evolved from an ethos that was indifferent to the needs of the poor and minorities. Such indifference in part has led to the evolution of organizational agendas that have either no clear concept of social ethics or with social ethics skewed in such a way as to make them appear almost callous.

However, as will become more evident from what follows in this and sev-eral subsequent chapters, no single factor can be identified as the prime culprit. The reasons for the agenda omissions are many and interrelated. To a lesser extent, the chapter also examines the impact of introducing race and similar social dynamics into current environmental policy consid-erations. But make no mistake: the issue is not just one of race. The larger issue examined is the reason for the absence of almost any concern in these agendas about the differentiated social/economic/health effects of both environmental activities and policies.

The chapter explores the idea that in many ways, concerns over social/economic/health conditions and the ramifications of environmental poli-cies and activities may have been simply beyond the threshold of percep-tion of these organizations. It is not a question of rejection, but rather one of obliviousness to the existence or importance of such social ethics issues in environmental policy. One policy analyst noted a lack of concern and indifference exhibited when exploring with a group of environmental sci-ence and policy students the economic ramifications of a blanket imposi-tion of a certain environmental policy that could have a large negative im-pact on regional employment and economic well-being. As he commented to one of the most vocal students, who felt nothing should stand in the way of such a "sound" environmental policy, "I can see you have never been hungry."

Chapter 4 follows up on the exploration of Chapter 3 with an examina-tion and argument for why the environmental-justice movement has only recently become such a public issue when the conditions of differentiated environmental impacts have existed for so long. It observes that all social and economic policy movements must go through an evolutionary process which for the nascent environmental-justice movement did not coincide with the mainstream environmental movement. This evolution has been described in different ways, but typically, such descriptions begin with an issue-recognition stage, followed by ever-increasing public awareness and debate on the issue, culminating in the formulation of specific legislative agenda items that may or may not be enacted.

The final two chapters in Part 1 consider first the question of minority attitudes toward environmental issues and then the problems of establishment agencies' attitudes toward minority or class concerns. Chapter 5 summarizes a series of national surveys conducted over a number of years on minority attitudes toward several dimensions of environmental policy and environmental conditions, and the relative importance of environmental protection versus other socioeconomic conditions, such as crime, employment, and education. These surveys reveal a level of minority concern about environmental issues not too dissimilar from that of the majority population. These surveys do not support the notion of minority lack of environmental involvement or presence in the decision-making process due to lack of environmental concern. What does emerge is a suggestion that in some cases, while important, environmental issues may, however, occupy a lower position of importance relative to issues such as crime, education, or poverty.

The last chapter in Part 1, Chapter 6, looks at the U.S. Environmental Protection Agency (EPA) as a representative of the general attitude that governmental environmental agencies have exhibited toward the issue of environmental justice. Both in personnel makeup and personnel attitudes, the EPA, in spite of a concerted effort to introduce environmental justice as an issue in their policy process, faces a difficult problem of cultural resistance. Even more to the point, the chapter points out that relative to most of the other environmental and natural resource management government agencies such as the Department of Energy or the Department of the Interior, the EPA has actually come the furthest.

One

Introduction

By one estimate, three out of every five African American house-holds currently live near a hazardous-material storage area.

Fines imposed on polluters by all levels of government in white communities in the 1980s were 46 percent higher than those imposed for violations in minority communities.

Fines levied against site violations under the federal hazardous-waste statutes were 500 percent higher in white communities than fines in minority communities.

Even after controlling for income, it takes 20 percent longer for a toxic-waste site in a minority community to be listed on the Super-fund National Priority List than a site in a white community.

Between 1990 and 1993, of more than 22 community grant requests to study the feasibility of locating a nuclear waste facility within a locale, 16 came from Indian reservations.

Until the early 1990s, the U.S. Environmental Protection Agency had conducted no major studies on the possible uneven distribution of environmental cost or benefits across racial or income categories.

The preceding observations raise disturbing questions: Are minority communities and individuals burdened with more than their share of environmental risks in this country, while enjoying fewer of the benefits of environmental regulation than others? Do these apparent inequities stem from racist or race-based policy? In other words, is environmental policy no different from education, criminal and civil justice, and a host of other socioeconomic institutions in this country in being tainted by the broad brush of race and class discrimination? If not, what besides race and class discrimination could possibly explain these differences in environmental burdens and benefits? Finally, and even more disturbing, what explains the apparent lack of concern for the uneven impact of environmental policies and activities in most of the original federal environmental legislation?

The Public Debate

Since the mid-1980s, there has been growing recognition that persons of color in both urban and rural areas may be exposed to much greater environmental risks than the American population as a whole.[1] These risks take several forms. Compared with the majority of the population, minority population groups in many cases live closer to facilities for the treatment or disposal of hazardous material.[2] They may further face disproportionately high exposure to other environmentally risky commercial and industrial operations.[3] Such communities, it appears, also may endure exposure to environmental hazards such as lead poisoning in their homes.[4]

Although this lack of environmental equity has existed for quite some time, only within the last few decades has the larger body politic begun either to recognize its existence or to place a higher priority on addressing it. Explanations for this neglect abound. Each major participant in the debate over environmental equity offers his or her own explanations. For example, those active in the grassroots environmental-justice movement argue that

> Such neglect exemplifies racially motivated decision-making, both public and private. In this "environmental racism," persons of color, either by design or neglect, endure a disproportionate share of the cost of economic development and growth, without enjoying a corresponding share of either the economic or environmental benefits. They have virtually no representation among the major nongovernmental environmental organizations and below-average representation within environmental agencies such as the U.S. Environmental Protection Agency (EPA).
>
> Minority populations, faced with few economic alternatives and not fully aware of the assumed risks involved, have grudgingly undertaken environmentally hazardous activities. The assumption of such "forced risks" results in many environmentally racist outcomes, such as disproportionate Hispanic exposure to farm pesticides due to their

overrepresentation as farm workers, or Native American communities' "choice" of potentially high-risk environmental activities like nuclear storage.

Minority populations are kept, often deliberately, from entering the environmental decision-making process by inadequate information or an incomplete understanding of the dynamics of the processes or conditions they confront. When major environmental risks are involved, these communities are thus unjustly placed at a tremendous disadvantage in deciding where to work or live.

Because of their relative lack of political and economic power, poor minority communities are frequently targeted for the location of environmentally hazardous activities. With everyone saying "not in my backyard," such activities get put in the backyards of those with the least ability to protest.

On the other hand, those who would question the validity of environmental racism's impact argue that

Environmental issues have historically held a lower priority within the minority community's political and economic agendas than such immediately pressing socioeconomic issues as education, drugs, crime, and unemployment. Thus, exposure to environmentally risky activities does not incur the opposition it would in majority communities. This attitude, rather than the lack of political or economic power, leads to the location of still more environmental hazards in minority communities. Furthermore, minority underrepresentation on various environmental action, planning, and policy bodies both in and out of government is a direct consequence of those communities' relative indifference to environmental problems.

Observed differences in environmental impact across communities or population groups are not racially motivated, but rather result from simple economic and physical factors operating in a free market. For example, hazardous-material disposal and treatment sites are simply located in areas where the cost of land use is low and the geology favorable. The fact that a disproportionately high percentage of minorities may reside near some of those sites is coincidental. Furthermore, many hazardous sites in urban areas are the remnants of earlier industrial activity that began when the communities were inhabited primarily by the majority population.

Exposure to environmental hazard or potential risk exposure is based not on racial bias but on differences in income or wealth. In general, the poor suffer from the forces of economics. Such forces "naturally" result in some people, despite race or ethnic origin, living in less desirable circumstances, including increased potential environmental risk. Environmental health hazards are just one component of a larger situation that includes higher incidences of crime, lower educational opportunities, and lower quality of health care. The same economic explanation applies to the disproportionately high exposure to lead poisoning and other environmental conditions associated with living in a decaying urban infrastructure or an economically depressed rural community. Given the high correlation between income and race, income bias could thus be misinterpreted as race bias.

In many cases of locational clustering of environmental risk, the hazardous sites existed before the minority population's influx into the area. Minorities choose such locations because they are affordable. In each case, the choice was an individual household decision, where increased risk potential was balanced against lower housing costs.

While not discrediting any of the above explanations of either side, the present work in part seeks to dispel the notion that any one explanation can account for all outcomes, conditions, and behaviors included under the widening umbrella of environmental justice. Furthermore, as later discussion will point out, insistence on a single explanation can lead to erroneous conclusions and misguided policy remedies at the national level.

Current nascent environmental-justice research has yielded little conclusive evidence for or against the above arguments. Much of the small but growing body of literature has focused on verifying, announcing, and highlighting that differences in exposure to environmental risk exist between minority and majority communities. In addition, while varying in analytic rigor, many of these studies, particularly those that evaluate risk exposure, have been hampered by the inadequacy of current environmental risk and demographic information. And the limited range of available methodologies has restricted most of these studies to considering no more than one potential environmental risk factor or issue at a time. Unfortunately, without a heroic manipulation of variables and assumptions, such handicapped research yields an incomplete picture of the total environmental differences across a community or a country.

Nor do we have the luxury of waiting for the development of "the good stuff"—that is, research methodologies that produce accurate assessments of all environmental conditions communities face. This is not an unusual circumstance in modern policy-making. In fact, it may be argued that no major policy decision has ever been made with the proper or sufficient analytic tools readily available. Whether employing good methodology or bad, whether using strong analysis or weak, whether the proper data are available or not, policy-makers are currently making major decisions about environmental justice in this country. And there is a cost for bad decisions.

As the historical evidence about any socioeconomic issue shows, bad or ill-advised choices made at the beginning of an effort take an extraordinary amount of energy and resources to correct.

Arguments of This Book

In light of this inadequacy of information, I encourage opportunities for better questions to be asked, and I offer a different perspective on aspects of the problem. Nevertheless, this book is not merely an exercise in academic voyeurism. Adopting the more analytical perspective I propose, however limited, reveals the following clear principles of policy behavior principles that will color all the following discussions:

Issues of environmental justice and racism involve far more than just the unequal exposure of minority or economically or politically disenfranchised groups to environmental health risk. At least as important are the interrelated problems of the reduced environmental quality of life some groups endure, the lack of participation of some communities in the environmental policy- and decision-making process, and the very serious asymmetry of governmental and private-sector response to the environmental concerns and demands of minority and low-income communities.

Expanding or redefining the "affected" group to include low-income populations or other disenfranchised groups does not appreciably change the problem of environmental justice or its solution. The current debate over whether race or income is the primary determinant of environmental injustices may have some significance in terms of statutory remedies, but from a policy perspective, this question is considerably less important.

The question of environmental justice in the United States must be considered in the context of the much larger but related international problem of environmental justice. Issues of national sovereignty, paternalism of developed nations, and economic necessity are all part of environmental justice on the world stage. Careful consideration must be given to any remedy that simply moves the problem to another country with no thought of its consequences to that nation's environmental health.

The preference for the prevention rather than the disposal of hazardous waste is understandable. The EPA's current policy has stated the long-term goal of waste prevention. But various forms of hazardous waste are still being generated and must still be disposed of. LULUs (locally undesirable land uses) will still remain, regardless of intentions. This reality will probably be with us for a very long time. Reducing the stream of waste should not be confused with eliminating it. Nor can a universal policy of NIMBY (not in my backyard)—or, as some put it, BANANA (build absolutely nothing anywhere near anything)—succeed today. Environmental-justice solutions that ignore the harsh reality of hazardous waste have an intrinsic flaw. They propose solutions that are either unworkable or unmanageable. There are certainly solutions that can change the pattern and method of facilities distribution. Hazardous-waste treatment facilities may be more evenly distributed, and there may be more equitable compensation for enduring such burdens. But the burdens cannot be eliminated. As long as it exists, the waste must end up in someone's backyard.

The New Paradigm

A major purpose of this study is to explore methods for evaluating environmental-justice problems within the context of public-policy analysis. At the same time, I also maintain that any such evaluation for a specific community is secondary to a larger concern. That larger concern is the need to accommodate differentiated environmental impacts in the policy-making process, to understand the differential burden various populations bear, and to devise means to accommodate all affected environmental stakeholders at the policy discussion table.

Stated another way, the principal mission of this work is to explain why environmental justice represents a new paradigm, why it represents the beginning of a major change in the very soul of the environmental movement in this country and possibly the world. In fact, expanding our base of comparison, the greatest and most challenging questions of environmental justice may not be due to inequity of environmental impacts in this country, but rather such inequities throughout the world.

Given the above argument, I would contend that a significant part of the current discussion about environmental justice is poorly directed. Discussions of what constitutes environmental justice should and will continue, but trying to reach specific conclusions in most cases will not change the larger picture and may be wasted effort. The larger issue is that a new and significant element has been introduced into the environmental policy process: For the first time in the environmental policy and action debate in this country, the differentiated impact of environmentally related private and public activities and policy in social, health, and economic arenas has become a serious issue.

But what is so startling about considering the differentiated impact of a particular public or private policy or activity across racial, ethnic, income, or any other category? Is this not the norm? Are not all public- and many private-sector policy decisions and activities examined under this light? In education, it would be considered absurd and irresponsible to implement a school-funding policy or a teaching policy that ignored the differentiated impact the policy may have on different income or racial groups. Indeed, the very reason for having formulas for the equalization of state school funding is an explicit recognition that all school districts and all students are not the same.[5] One of the first principles of public finance is the idea of vertical equity, in which taxes are evaluated on how they affect people in different classes of income or socioeconomic conditions. Likewise, a major component of criminal justice policy in this country is addressing the problem of unequal justice and protection across racial and income categories. And can anyone conceive of a national health care or housing policy that ignores differentiated race or income impacts?

Yet the possibility that environmental regulations, policies, agendas, and private activities may burden or benefit some racial or income groups more than others simply has not been a major concern of the environmental movement or of the policy agenda of this country. Nor is it even a question of agents, during the implementation of a program, ignoring a legislative mandate that included a charge of environmental impact equity or justice concern. Until the last several years, such mandates, at both the federal and state levels, did not exist. At the federal level, the major environmental enabling legislation includes no indications or charges that race or income effects are important. There is, in fact, almost a denial that environmental policy legislation and activities have a social impact.

Ground is only slowly being given on this issue by more traditional mainstream environmental activists. Marching under the banner of objective evaluation, these activists appear to devote more effort to challenging any finding of environmental racism or environmental injustice than to moving on to the larger issues, such as considering the differentiated impacts of environmental policy and activity as a matter of routine. The tone of some prominent voices both in and out of government is almost one of indignation that anyone would try to interpret environmental policy actions and outcomes in terms of such considerations as race and income. Many environmentalists prefer to think about all of humanity rather than descend into mundane particulars about specific racial groups, income strata, or geographic regions.

The German sociologist Ulrich Beck, in *Risk Society,* one of the best-known works on the universality of environmental risk today, argues that the strongest feature of modern society is that everyone, regardless of race, income, or ethnic class, faces similarly high levels of risk (what he calls "the worldwide equalization of risk positions"). Beck recognizes that historically, factors such as income and class have had a significant mitigating effect on the impact of environmental risk, but he counters that in today's society, the more significant reality is that no one can escape the impact of that risk. At the same time, Beck also points out that "there is a systematic 'attraction' between extreme poverty and extreme risk. In the shunting yard where risks are distributed, stations in 'underdeveloped provincial holes' enjoy special popularity."[6] Again, what we may have here is a useless debate. For purposes of environmental-justice policy, it is not necessary to disprove that the majority of environmental risks are faced by all. Nor is it necessary to recognize that some significant portion (not even the majority) is unequally distributed. If some significant portion of that risk is unequally distributed, perhaps only 20 percent, then does not that unequal impact deserve specific policy attention?

The capacity for self-deception and convoluted thinking is apparently still alive and well within the environmental movement. Hardest of all for some in the environmental movement to accept is that environmental racism is more than just a catchy phrase—that it is possibly a real phenomenon that has influenced some public and, especially, private environmental decisions and policies. So reluctant are many to accept the possibility of environmental racism that almost any other explanation for differentiated outcomes appears more attractive. Critics of the specific charge of environmental racism in policies or activities thus curiously leap to embrace a given study that suggests that economic need was the primary determinant of a particular environmental event, such as the placement of a hazardous-material facility. These defenders of the status quo accept that environmental discrimination does occur, but they are more able, or perhaps more willing, to accept that it is based on class or income rather than race. This stance ignores that differentiated environmental impacts across income classes represent the same problem of the neglect of socioeconomic impact by environmental policy as do differentiated impacts across racial groups.

An Outline

Before discussing the problem of environmental justice, we must first establish its parameters. They include the problem of creating workable definitions of the major terms of the discussion. From workable definitions, the discussion turns to an examination of the development and evolution of the environmental-justice movement and its relationship to other issues or currents of public policy. This includes the larger mainstream environmental movement, the civil rights movement, and the even larger area of public policy on social impact. After establishing the "what," we must explore the "what next." That exercise in turn will include an outline of the basic problems of measurement facing any analysis in this area and some simple guidelines for any analytic examination. Because of the considerable misunderstanding and misconception about attitudes, several chapters will treat the question of community attitudes toward environmental questions, and the policy and attitudinal problems, both in and out of government, associated with mainstream environmentalism's response to the challenge of environmental justice. Also included will be an exposition of the basic principles of public-policy analysis that could guide an examination of this issue and the formulation of any solutions.

Two

Forms of Environmental Justice

In the current discussion on the socioeconomic and health impacts of environmental activities and policies on specific communities, especially minority communities, an extraordinary amount of time and energy is wasted on debate over labels and definitions. The worst waste of time occurs because parties in the discussion don't understand that others in the debate assign different definitions to the same terms.

Definitions

In any public-policy context, definitions are important. In the discussion on unequal effects of environmental policies or activities, familiar words such as *racism, equity,* and *justice* have been appropriated for use in the ensuing debate. Given that the movement itself is young and still plagued by misunderstanding, the uses of these three terms—*racism, equity,* and *justice*—in this context are also young and unclear. Misunderstanding and misinterpretation abound. In the interest of clearer dialogue, I offer the following definitions for policy analysts. However, these definitions should not be construed as definitive. I have tried to formulate a working definition of each term without an exhaustive defense or exploration of its myriad nuances or implications.

In addition to defining the three major terms used for describing environmental impact differences, the chapter also offers a system for classifying the various inequitable situations. Misunderstandings about the varying nature of environmental impact situations have often impeded resolution.

13

One size does not fit all, and one policy prescription does not fit all environmental impact conditions. For example, a problem may be both geographic and population specific, such as a specific minority community in a specific city living near a specific hazardous-waste site. Or it may be population specific without regard to location, as when migrant Hispanic farm workers are exposed to pesticides at a disproportionately higher rate than the general population. Furthermore, the differentiated impact may be based not on population or location, but rather on economics, as when lower-income families are unequally exposed to lead poisoning.

Classification also permits us to consider the merits of including one particular condition under the general umbrella of environmental justice without challenging the validity of the whole. For example, in speaking of environmental justice, many cite the problem of above-average subsistence fishing by certain minority communities with an accompanying increase in exposure to toxins in the fish. However, others, recognizing the voluntary nature of the act, challenge whether this behavior and the accompanying increased risk exposure really come under the rubric of environmental injustice. Within our classification system, such subsistence fishing is a "non–area-specific, economics-based" condition that may or may not be a valid environmental-justice issue.

Environmental Racism

I begin with the phrase that seems to capture the most attention and carry the greatest emotional impact. The term *environmental racism* was introduced in the mid-1980s by several sources, such as the United Church of Christ (UCC) in their 1987 report *Toxic Wastes and Race in the United States: A National Report on the Racial and Socio-Economic Characteristics of Communities with Hazardous Waste Sites.* In that report, the term *environmental racism* indicates the disproportionate environmental burdens imposed on groups or communities as a result of their minority status. Definition of the term was later expanded to cover the exclusion of representatives of such racial minority groups from participation in the formulation of environmental policy.

The word *racism* carries considerable emotional baggage. Many employ it as a synonym for any kind of inequality that affects minorities. They argue that racism occurs, regardless of original intent, if the action results in consequences that are more adverse for minority communities than for other groups. Many others hesitate to use this potent label unless the actions described are both deliberate and systematic. They counter that describing all such actions as racism dilutes the term's impact. One of the most useful definitions that recognizes the potency of the term comes from the Racial Justice Working Group of the National Council of Churches:

> Racism is racial prejudice plus power. Racism is the intentional or unintentional use of power to isolate, separate, and exploit others. This use of power is based on racial characteristics. Racism confers certain privileges on and defends the dominant group, which in turn sustains and perpetuates racism. Both consciously and unconsciously racism is enforced and maintained by the legal, cultural, religious, educational, economic, political, environmental, and military institutions of societies. Racism is more than just a personal attitude; it is the institutionalized form of that attitude.[1]

By this standard, while any population could be guilty of racial prejudice, only those capable of institutionalizing such racial prejudice could be guilty of racism. Obviously, some would debate this distinction, arguing that anyone can practice racism as long as any negative effects are based on race. By this latter, more expansive, argument, a group practicing racism need not be dominant within the larger society; it need only be capable of institutionalizing its racial prejudice within some sphere of society.

Much of the debate in this environmental area, even some of the current scholarly work in environmental justice, is more concerned with the naming of things than the solving of problems. However, some policy-makers and analysts would argue that without careful naming, a proper solution cannot be found. (In other words, without forcing society to perceive certain events as examples of racism, solutions will address only the symptoms of racism rather than its root causes.) Ultimately, this argument may be correct. From another perspective, however, insisting on precise and specific labels for those environmental conditions invites a debate that, although appropriate, may actually slow the solution of the problem.

From the standpoint of environmental policy formulation, one might properly ask, How important is it to prove racism? First, a charge of racism may call into action a completely different set of statutory remedies than those that would apply to a simple charge of inequity. The charge of racism may also carry special ethical weight in the formulation of remedies. But from the perspective of policy analysis, the charge carries less significance. Policy-makers are usually less interested in identifying villains than in finding a solution to existing problems and preventing future occurrences.

Furthermore, precisely defining environmental racism is not easy. At one extreme, environmental racism describes only those circumstances in which race was an intentional and deciding factor. At the other extreme, environmental racism means any condition or policy in which people of a specific racial group are disproportionately affected. Many of the circumstances discussed in this book fall under this broader definition. Conversely, it would be extremely difficult to prove occurrences of the narrower definition, which requires explicit intent.

From a policy analysis perspective, the narrower definition may be the more workable one. Even though the race of those affected may be crucial to some individual instances of environmental inequity, it does not follow that every negative environmental impact on minority populations is intentionally racist. Furthermore, broad application of the label of environmental racism to all negative outcomes not only hinders the formulation of policy but dilutes the significance of those situations in which racism was intentional.

Environmental Equity and Inequity

The term *equity*, or perhaps more properly *inequity*, carries considerably less emotional baggage than *racism*. A term used in slightly different contexts in different social sciences, *inequity* usually denotes differences across groups along a particular dimension, such as income, race, education, location, or occupation. When most people use the term *inequity*, they are not implying that such an outcome is the result of an intent to harm or exploit. The term only indicates outcomes, not intent.

In economics, for example, equity is usually depicted as an alternative to the efficient use of resources.[2] Two decision-makers thus must often decide, on the basis of some variable principle of equity, between efficient use of available resources (including income) and achieving a higher level of equity. For example, adopting the principle that no one should live below the poverty level may mean diverting resources from their most efficient use. Conversely, focusing exclusively on efficient resource utilization could mean ignoring the plight of many of the poor.[3]

The application of these microeconomic concepts of equity to environmental conditions, policy, and outcomes would require a very complex analysis.[4] The principles of equity developed under this conceptualization apply not only to individuals but to other agents in the environmental process, such as private firms, governmental entities, and owners of resources. Such a system of equity can be applied to issues as varied as rates of environmental resource usage, physical and nonphysical property rights, and methods of taxation and environmental finance.

The term *environmental equity*, like *environmental racism* and *environmental justice*, has been in use only a few years. Just as *environmental racism* was coined to describe conditions that reform groups were fighting against, *environmental equity* was adopted to describe an ideal or objective toward which groups were striving. One EPA report frames environmental equity in the context of current environmental protection and regulation policy: it "is the equal protection from environmental hazards of individuals, groups, or communities regardless of race, ethnicity, or economic status."[5]

Within the environmental policy venue described by the EPA definition, the definition of *environmental equity* is much more narrow than the more generalized economic definition of equity described above. This environmental definition implies a reaction to equalize or protect after the fact and focuses only on negative environmental impacts. Common usage may expand the definition of environmental equity to one that is more proactive, thus addressing the distribution of environmental benefits. But even with these modifications, environmental equity still ignores participation by all who have a stake in the environmental policy-making process, as well as those helping to define the actions needed to remedy present inequities.

Reflecting this refinement in environmental policy, the term *environmental justice* has for some supplanted *environmental equity* as the movement has evolved. This more expansive and proactive phrase has become a label of convenience to identify concepts or conditions for which labels did not previously exist. Of course, there is nothing sacred about the current definitions except that they are convenient and generally agreed upon. To add to the confusion, in the current literature, the phrase *environmental equity* is used as a synonym for conditions also described by both *environmental racism* and *environmental justice*.

Environmental Justice

Environmental justice, like the terms *environmental racism* and *environmental equity*, invites debates over its definition. Environmental groups, civil rights organizations, and numerous community-action groups have developed their own definition of the term. For some, it is simply a synonym for the same circumstances or outcomes that they also may label as environmental racism or equity. Even within the EPA, the staffs at each of several administrative levels and in various programs are developing their own definitions of the term.

Discussions on environmental justice cover a wide range of both conceptual and operational issues, including such questions as: Should extraordinary measures be adopted to correct past wrongs, or should attention instead be focused on preventing future injustice? Most agree that agents, both inside and outside government, should give some attention both to correcting past wrongs (particularly those that have created present dangers) and to working to prevent future injustice. But in a world where resources are scarce, remedies are not limitless.

Does environmental justice imply that all affected classes of peoples should experience equal environmental burdens and benefits? Or does environmental justice mean that every affected class has an equal voice in the distribution of those burdens and benefits (resulting in a distribution of burdens and benefits that may or may not be equal)? Viewing environmental justice from one of these two perspectives can result in radically different outcomes. Furthermore, both have drawbacks. Adopting the viewpoint of the first necessitates the adoption of an extraordinarily complex and most certainly costly system of measuring, evaluating, and subsequently redistributing burdens and benefits. Actually, much of such a task is well beyond the methodology and administrative resources currently available at any level of decision-making. Following the second interpretation leads to issues of political empowerment and participation. This second approach also makes a social policy decision. It sidesteps the equal distribution of effects challenge by focusing on how and by whom decisions are made.

Putting aside questions of the precise intent of environmental justice, we are still left with the equally contentious issue of which groups or populations to include under the environmental-justice umbrella. Should income or regional location count as much (if at all) as race or ethnicity in environmental-justice schemes? Under the tenets of environmental justice, who is entitled to redress? Who should have standing under the environmental-justice umbrella? And just as important, who should not have standing?

From a policy analysis perspective, redefining the "affected" group to include economically or politically disenfranchised groups, regardless of race, does not appreciably change the basic problem of environmental-justice or policy solutions. Of course, procedurally, the differences between the race- and income-based definitions do influence which current statutes we may choose as remedies.

Favoring the most expansive definition of affected groups places environmental justice within a larger social contract. That contract implies that all individuals should have equal access to environmental protection and equal opportunity to enjoy environmental benefits. Equal access, however, does not imply precisely equal protection, only the absence of unequal impediments to protection. Conversely, limiting the contract implies that excluded population classifications, groups, or individuals do not merit environmental justice. So stated, this position would be hard to defend.

During the environmental movement's brief existence, the concept of environmental justice, like those of environmental equity and environmental racism, has already taken on many shades of meaning. The bad news is that each of these meanings carries with it certain operational handicaps. The good news is that the following unofficial definition of *environmental justice,* as worked out by the EPA's Environmental Justice Office in consultation with several study and advisory groups, provides an excellent first step in framing an operational interpretation:

> Environmental Justice: The fair treatment of all races, cultures, incomes, and educational levels with respect to the development, implementation, and enforcement of environmental laws, regulations, and policies. Fair treatment implies that no population of people should be forced to shoulder a disproportionate share of the negative environmental impacts of pollution or environmental hazards due to lack of political or economic strength.[6]

I would add only the following phrase after the words "pollution or environmental hazards" in the last sentence: "or be denied a proportionate share of the positive benefits of environmental regulation or program."

The EPA definition is a good beginning, one that should guide our development of an operational definition:

- The EPA definition speaks of *fair*, as opposed to perfectly equal, treatment or protection. Equal distribution of negative or positive consequences implies perfectly homogeneous preferences and circumstances, which have no validity in the real world. Precisely equal treatment is thus an impossible operational standard.
- The EPA definition also allows that disproportionate effects *can occur*, but they are unfair only when they result from a lack of political or economic power, as opposed to resulting from well-informed free choice.
- Unlike the term *environmental racism*, the term *environmental justice*, as defined by the EPA, includes people as identifiable groups rather than as racial distinctions. Operationally, this means that time and effort are not wasted in meaningless debates over which groups to include under the broad umbrella of environmental justice.[7]

Of the three major terms describing unequal environmental treatment and protection—*environmental racism, environmental equity,* and *environmental justice*—the one used most often in this work will be the most inclusive: *environmental justice.* This choice does not, however, imply a diminution of the other two terms.

Measuring Environmental Impact:
An Operational Definition of
Environmental Justice

Whatever the outcome of the philosophical debate, a serious operational problem remains. Indeed, an operational definition of environmental justice is needed perhaps even more than a philosophical or conceptual one.

Although the topic of evaluating and implementing environmental justice will be developed more thoroughly in Chapter 8, I offer a few general comments here.

One of the most controversial issues in the current operational debate revolves around the proper level of geographic disaggregation for evaluation. The choice of the level of geographic evaluation plays a prominent role in arguments over whether an environmental-justice imbalance actually exists—either nationally or in specific locales.[8] Studies champion either counties or postal zip code areas, or census tracts (or even smaller units) as the proper geographic units for evaluation.[9] This raises the question, If changing the geographic unit can change the results of a study, how valid are those results?[10]

Similarly, debate occurs over the level at which a community is classified as minority or low income. Some studies define a minority community as one in which minorities make up 50 percent or more of the population. Others define a minority community as one where the minority population percentage exceeds city, state, or regional averages. According to the second definition, a community with only 9 percent African Americans could be classified as a minority community if the city in which it exists had a minority population of 6 percent. Conversely, a community with a minority population of 55 percent would not be considered a minority community if it was within a city with a 70 percent minority population (although it would be classified as a minority community under the first definition).

Yet another approach to community definition for environmental-justice purposes makes no explicit attempt to identify communities as minority. In this approach, minority percentage is used as a continuous variable. Thus, typically, the investigation would not focus on contrasting conditions in minority versus majority communities, but rather on identifying changes in community environmental conditions that occur when minority percentage changes.

The assessment of environmental justice is particularly challenging for the EPA and other federal agencies with environmental responsibilities.[11] Throughout their history, these regulatory agencies have focused almost exclusively on problems and solutions that are easier to define. For example, is a particular chemical present in a given environment? Does the level of a particular pollutant in the ambient air exceed an established threshold? Has a particular level of compliance with environmental standards been achieved? While there can be debate over whether a given threshold for a particular environmental condition is appropriate, the simple existence of a chemical or the scale of its measurement can be objectively observed.

The debate in environmental justice, however, is at a much more fundamental and fuzzy level. Environmental justice deals as much with the so-

cioeconomic impact of a phenomenon as with its chemical or biological impact. Unfortunately, the EPA—and to an even greater extent, the other federal agencies with environmental responsibilities—has little experience in handling such blended chemical-biological and socioeconomic problems.

The Single-Risk Assessment

Yet another operational problem is related to risk assessment. Ironically, those who insist on establishing rigid definitions of environmental-justice conditions generally choose an inappropriate analytic methodology. From the standpoint of policy analysis, the typical risk methodology, while often appropriate for single-source risk assessment, is ill suited to evaluate the multidimensional problems usually found in environmental justice.[12] Well-defined measurements and constructs are thus input into poorly prepared algorithms, with ill-defined results.

Much of the endless discussion in evaluation methodology stems from a conflict between two different approaches. One seeks rigid, well-defined measurements and standards of evaluation. The other relies on more flexible and fuzzy standards. Consider the example of the EPA and its nongovernmental supporting groups and organizations. Historically, the EPA has favored a more rigid approach. Unfortunately, the EPA's mission and reputation have suffered from this early (and still favored) tendency to establish or seek fixed universal standards. This approach is consistent with the ineffective command-and-control approach to regulation.[13] And whether it is a reflection of an "engineering" mind-set or other organizational maladies, this approach has caused the EPA almost as many problems as it sought to resolve, both in policy implementation and goal attainment.

If the EPA adopts this same attitude in dealing with issues of environmental justice or is saddled with legislation that has such command-and-control specifications, environmental-justice solutions will suffer. This is because, as countless studies have shown, the rigid approach allows little room for the situational variations encountered in the real world. Absent such flexibility, program officers spend more time chasing illusory target thresholds or performing set protocols than addressing the problems of a specific community. Such a rigid focus seldom leads to a solution. More often, it leads to needless debate about the validity of levels.

The flexible approach to environmental regulation focuses on performance principles or standards, not specific threshold levels. In many cases, the environmental injustice may be one of comparative disadvantage. For example, Community A may face greater environmental risks than Community B. But the risks in both communities fall below some absolute

threshold, and under a rigid standard approach, no remediation would be called for. In such a case, however, the flexible approach to environmental regulation, unlike the fixed-level approach, might still find cause for environmental-justice concern. The flexible approach would allow the performance principle of relative disadvantage to still point to a need for remedial action.

A relativistic regulatory approach changes the focuses of concern from thresholds and conditions in the environment to comparative environmental conditions of communities or other groups. This flexible approach has the possible disadvantage of not being very conducive to centralized control. A centralized authority can establish the guiding performance principles in a flexible environmental regulation approach. After that, however, most of the actual decision-making authority rests with agents far below a central governing authority. In this approach, the role of the central authority shifts from controlled management to (at best) a referee of disputes. Unfortunately, in many cases, some policy-makers simply do not have enough faith in agents outside a central authority to believe that such a decentralized method of operation can work.

Forms of Environmental Justice

To better understand the environmental-justice issue in its many manifestations, I have identified several broad classes of problems. These categories combine traits from both the rigid and flexible approaches to environmental evaluation and regulation. The examples below, ranging from the obvious to the subtle, are intended to show which situations have environmental-justice dimensions, not which are definitive examples of environmental injustice. They involve problems in both environmental outcomes and the implementation of policy. Two key characteristics divide the categories: specificity of location and time frame of the decision process. These two sets of traits can be formed into the two-dimensional matrix described in Table 2.1.

In Table 2.1, note the distinction between problems that arose in the past and those whose source is in the present. From a policy perspective, in many cases, expending huge resources in order to assign "ultimate" responsibility for conditions that arose in the past has little merit. This apparent lack of concern over assignment of responsibility does not excuse past excesses or dismiss the ethical questions raised by past actions or policies. Nor does it ignore the fact that certain outcomes may have directly resulted from "normal" market forces, and that policies designed to change such outcomes face the extremely difficult task of fighting those "natural" forces. Rather, this position argues that for the development of remedial

Table 2.1. Environmental justice categories.

Area	Historic condition	Current or future condition
Single location or area specific	Inequities result from past location decisions in a given area	Inequities result from a current of future location decision in a given area
Multiple areas	Inequities result from a historic pattern of decisions or policies across a collection of areas	Inequities result from a current or future decision or policy affecting a collection of location-specific conditions
Non-area specific, population based	Inequities result from past decisions or policies affecting a population	Inequities result from a present and still valid decision or policy affecting a population
Non-area specific, economics based	Inequities result from a historic economic need of a population for which the present population may suffer	Inequities result from a present of future economic need of a population

policies, it may be unimportant whether an action was premeditated or was the indirect consequence of the pursuit of other goals.

For example, the location of hazardous-waste landfills in urban minority communities may represent the bias of some industries for leaving wastes in communities that traditionally possess little political power. However, this pattern could also simply represent the historic preference of heavy industries, which were primarily located in urban areas, to leave their waste near their operating facilities—which, having been abandoned, were later surrounded by arriving minorities.[14] Whatever the explanation, the problem is not why industry took the action it did, but that minority populations now living and working in those areas may face a greater exposure to hazardous materials than the rest of the population. And the important policy issue is what to do next.

Whether the past decisions from which present conditions arose were intentionally or unintentionally biased, the policy response most appropriately focuses on addressing the present exposure situation and not on determining past intent. Endless debates about the assignment of responsibility may too quickly lead to useless investigation of fuzzy historic events that will have little effect on what to do for present community risk.

A different policy perspective emerges when considering current or proposed activities with environmental-justice dimensions. For such ongoing or future activities, the issue of responsibility assessment takes on much greater significance. Policy decisions affecting present or future decisions about facility location or environmental quality are aimed at stopping an ongoing activity or changing a future activity. The policy focus is on prevention and modification. In such situations, assessment of responsibility has a direct bearing on policy.

For example, a proposed decision to locate a hazardous materials processing facility in a minority community, taking advantage of that community's relative lack of political and economic power, has serious policy implications. The company involved may actually be in violation of Title VI of the Civil Rights Act. Similarly, a firm's plans to locate such a facility near a minority community, regardless of its reasons, may still raise policy concerns—focusing on the adequacy of community awareness of the health and safety implications of the location decision.

In both the above examples, the policy focuses for a current or future operation condition will be very different from those of a policy concerned with present conditions arising from an operation that is no longer functioning. A part of the confusion in current environmental-justice discussion on community conditions is confusion between these two different situations. As will be explored later in this book, a single policy solution does not fit all situations.

Environmental Injustice at Specific Sites

One of the largest classes of environmental-justice problems is associated with a specific geographic community. In such a situation, a city, county, or neighborhood—usually inhabited by a minority or low-income population—faces a greater environmental burden than that faced by the population in the larger surrounding area. A historic example would be a community traversed by more major roadways—and thus plagued by more auto emissions—than any other community in the area. A present or future example would be a community chosen as a future site for an environmentally risky facility on the basis of the community's racial or economic demographics.

Past Injustices at Specific Sites

Along the lower Mississippi River, from Baton Rouge to New Orleans, is a complex mix of industries and residential housing, dominated by petrochemical facilities and African American communities, called Cancer Alley. The percentage of minorities living within two miles of one of these facilities is higher than for either Louisiana as a whole or for the entire

United States.[15] Jim Motaualli, in "Toxic Targets," writes, "The average American is subject to 10 pounds of toxic chemical release per year. The average convent (a predominantly African American Louisiana town along Cancer Alley) is exposed to 4,517 pounds."[16] Many communities along this corridor predated the establishment of the various industrial facilities that now appear to threaten them, so individual location choices were not made with awareness of the risks.[17] Although there is a higher incidence of cancer along this eighty-five-mile corridor than in the general population, it has yet to be linked conclusively to living near the facilities. Better demographic and environmental data as well as improved methodologies for risk assessment are needed. Nevertheless, anecdotal evidence strongly suggests such a link.

Another example dates from the late 1940s to early 1970s but still has repercussions today. The old Atomic Energy Commission, responding to a rising demand for uranium to fuel a growing nuclear power and weapons industry, opened hundreds of uranium mines in Cove and Red Valley, Arizona, and recruited more than fifteen hundred Navajo men from the area to operate them. The danger of radiation exposure during the mining process was known as early as 1949, and the commission's own health investigators predicted that the then current policies would result in an epidemic of lung cancer among the miners. Nevertheless, the federal government made no effort to reduce the risk or to warn the miners of the danger until the late 1960s. Federal agencies justified this policy as necessary for national security and for meeting the pressing need for uranium. As of the early 1990s, 230 of those Navajo miners are dead—most from either cancer or respiratory disease. (Compensation from subsequent programs has been extremely difficult to collect, primarily because documentation requirements show little sensitivity to tribal customs.) As an added environmental burden, the mining area, once an attractive natural resource, has been left scarred and barren.[18]

Finally, Atgeld Gardens, a Chicago public housing project for about ten thousand African American people, was built in what has been termed a toxic donut. Constructed directly over a landfill, the project is surrounded on all sides by hazardous-material facilities—including seven other landfills, a chemical plant, a paint factory, two steel mills, and a sludge-processing facility—most built after the housing development. Cancer, respiratory ailments, and birth deformities are significantly higher in this area than in the rest of Chicago, even in other predominantly African American communities. Evidence suggests that both city and Cook County officials were well aware of the environmental problems in Atgeld Gardens for several years but until recently had made no serious attempt to remedy them.[19]

Present and Future Threats to Specific Sites

In the early 1990s, on the two-thousand-member Lower Bule reservation of the Teton Lakota in South Dakota, the chair of the tribal council wanted to accept $4.5 million a year from the South Dakota Disposal System. This "generous" offer would allow the Denver-based waste management company to sell to the community for $1.00 land that the company currently owns in another part of South Dakota and on which it plans to operate a landfill that would serve several states. By manipulating the land to fall under Indian jurisdiction, the company evades much of the regulatory power of both state and federal systems.

Furthermore, during the same period, the Teton Lakota tribal council, along with fourteen other tribal councils, applied for several small ($100,000) grants to study the feasibility of serving as a short-term (forty years) storage facility for spent radioactive fuel from nuclear power facilities around the country. Despite the potential environmental risks, economic desperation and a dearth of alternatives for economic development have made both these proposals tempting for many tribal council members.[20]

Injustice in Multiple Areas

In the current debate over environmental justice, problems specific to multiple areas attract perhaps the most attention. In 1987, a study by the UCC brought the problem of differentiated environmental burdens to the nation's attention.[21] Starting from that early work, studies using counties, zip codes, and census tracts have attempted to document the possible differences between minority and majority communities as locations for hazardous materials facilities.[22]

Past Patterns for Multiple Areas

The original UCC study of 1987, which used the five-digit postal zip codes to define "communities," found that nationally, there was a significant correlation between race and the location of commercial facilities for the disposal of hazardous waste. The study found that even after controlling for income, regional effects, and urbanization, race remained a significant explanatory variable for the location of such facilities. Furthermore, the differences in racial composition were greatest between communities containing two or more facilities (or the largest facilities) and communities without any facilities. The former communities had an average minority population of 38 percent, the latter an average of 13 percent.[23]

Benjamin Goldman, using the county as the base unit, found that minorities have a 60 percent higher probability of living in counties that ranked in the top 2 percent for concentration of twenty-four industrial hazards. According to his research, African Americans had an 85 percent higher probability, Latinos a 32 percent higher probability, and Asian Americans a 300 percent higher probability than whites of living in those counties.[24] In a similar study in 1992, Leslie A. Nieves investigated four thousand industrial facilities whose production posed environmental hazards to air and groundwater, and emitted potentially hazardous radiation.[25] By using the county as the measurement unit, Nieves found a significant correlation between level of potential risk or hazards and percentage of minorities.[26] Neither Goldman nor Nieves found a similar significant correlation between location of hazards and income.[27]

Present and Future Problems

In the management of the Superfund toxic-waste cleanup, the EPA's performance and administration of program policies appear to differ significantly across communities. As of 1992, for the nearly twelve hundred sites on the Superfund or National Priority List (NPL), penalties under various hazardous-waste laws were 500 percent higher for sites with the greatest percentage of white population than for those with the greatest percentage of minorities. Minority community sites took 20 percent longer to be placed on the NPL than sites in white areas. In several EPA regions, comparing just those sites that make the NPL, it takes from 12 to 42 percent longer for the cleanup to begin in minority community sites. Further, in minority communities, containment is selected as the clean-up method 7 percent more often than permanent treatment. In white communities, permanent treatment was chosen as the clean-up method 22 percent more often than containment.[28]

Non-Area-Specific, Population-Based Justice Problems

Many environmental-justice problems are not specific to location but rather to a minority population's lifestyle or economic condition (or a combination of both).[29] This disproportionate exposure to risk has received less attention than problems related to location. Because they are often spread throughout a population, these non-area-specific conditions are far less likely to find an organized constituency and receive less attention than location-specific problems. In fact, some outside the environmental-justice movement even find it difficult to call such issues environmental injustice and ascribe the resultant conditions to the freely chosen lifestyle of individuals or the community.

Risk Exposure through Diet

Several minority groups—among them African Americans, Native Americans, and Asian Americans—consume relatively large amounts of fish. In a study of fishing in Michigan, some African American and Native American groups consumed more than five times as much fish as the amount on which Michigan levels of toxicity in fish were calculated. In Michigan, for example, acceptable levels of toxicity in fish are based on an individual consumption of 6.5 grams per day. One study, however, found that some African Americans ate 20.3 grams of fish per day and that some Native Americans between thirty and fifty years of age ate 30.[30] This is both a historic pattern and a continuing condition.

Lead Exposure

Minorities, especially African Americans, suffer from a much higher level of lead poisoning than the U.S. population as a whole, due to both living conditions and occupational hazards.[31] This is a historic situation based on economic conditions. In spite of the current ban on lead-pigment paint and lead plumbing fixtures, lead poisoning poses a deadly threat to many minority populations. In part because of the concentration of many minorities in older urban areas and in part because of the lack of preventive maintenance in such dwellings, minority children are much more likely than whites to suffer from the effects of lead poisoning, effects that include mental retardation, physical impairment, and even death.[32]

Pesticide Exposure

In California, between 80 and 90 percent of workers on nonfamily farms are Hispanic.[33] Nationally, 77 percent of the nonfamily member farm workers identified themselves as minorities. Of those, approximately 61 percent were Mexican and 11 percent were Latinos born in the United States.[34] This mobile workforce faces exposure to a wide variety of farm pesticides. Unfortunately, acceptable levels of pesticide exposure have been based on risk to the consumer and have not considered farm workers' immediate contact with such chemicals. Nor do they take into account the possible cumulative or synergistic effects of exposure to an array of pesticides. To compound the problem, migrant farm workers move from one farm to another, often across several state boundaries, and thus have virtually no health care service and no political constituency to support them. In only eight states, for example, are migrant farm workers fully covered under workers' compensation insurance.[35]

Economic need and a lack of education leave this group unable either to refuse working under such conditions or even to fully appreciate the magnitude of the environmental risks they face. But lack of good epidemiological studies on farm worker exposure and the synergistic effects of other lifestyle factors, such as higher rates of cigarette smoking and generally poor health care, make it extremely difficult to prove that pesticide exposure is the cause of all their health problems.

The Significance of Environmental-Justice Issues

Some may question the validity of the environmental-justice issues in the examples cited above. The case of the Navajo miners, for instance, could be interpreted as an example of government secrecy or bureaucratic insensitivity. There are many methodological flaws in most of the major national studies on differences between white and minority communities in proximity to dangerous sites. Some cases may be ascribed to reasons other than racism, such as the attraction of cheap land or access to transportation arteries. And it could be argued that the case of fish toxicity in Michigan is more a problem of public health than of environmental justice. Finally, some would claim that the Native American tribes that have offered their lands as waste disposal sites—since they did so freely and not as the result of any government or private duplicity—are not the victims of environmental injustice. But environmental injustice occurs whenever a community or a people experiences a greater environmental burden than that of the majority population. It does not matter whether these burdens were assumed voluntarily, whether equalizing compensation was given, or whether the problem could be better addressed via another public-policy venue, such as public health policy.

As noted earlier, I have provided these examples to illustrate the potential range and form of environmental-justice issues, not to provide definitive categories. In the same way, this work will not attempt to decide on the validity of charges of environmental injustice in each example I give. The important issue here is recognizing that conditions or circumstances like the ones cited are where the environmental-justice debate will be conducted. Environmental justice is an umbrella that spreads over past, present, and future, covering areas as diverse as public health and the economic development of communities.

Three

What Has Gone Before
Why Race Was Not on the Original
Environmental Agenda

The mainstream of the environmental movement—both governmental and nongovernmental—cannot successfully address the question of environmental justice without making significant changes in its attitudes. Among those attitudes is a reluctance to recognize the systemic biases embedded in the very foundations of the environmental movement. However, because the current environmental establishment developed from a foundation of conservation, it is not surprising that virtually all mainstream environmental organizations until very recently have ignored issues of environmental justice.

Until very recently, certain issues have been curiously absent from the agendas of mainstream environmental organizations. These issues include the unequal distribution both of environmental hazards and of the beneficial impacts of environmental regulation and policy across subsets of the population. More generally, until the last few years, most mainstream environmental organizations simply had no place on their agenda for the socioeconomic impact of environmental decision-making.

As observed by social justice critics, mainstream environmental organizations have tended to focus on things, rather than people.[1] People have been treated almost as a homogeneous mass: if one benefits, all benefit. To judge by the literature of mainstream environmental organizations, there are no poor or rich, no black or white, just polluters and defenders, land or fauna to protect, a single, generic humankind to consider. Overwhelmed by the forest, these organizations historically appeared to fail to see the trees.

The idea that some may benefit much more than others from specific environmental policies, and that those others may have a legitimate grievance, appears to have been an alien concept. Even more alien is the idea that such unequal distribution of burdens and benefits might actually be a legitimate source of concern for these organizations.[2]

The environmental movement, both in and out of government, is primarily white and to a large extent indifferent to issues of social justice. Minorities are virtually absent from mainstream environmental organizations. As noted by Donald Snow of the Conservation Fund, in his survey on the challenge to the leadership of the environmental movement,

> Practically none of the mainstream conservation environmental groups in the United States—regardless of location, scope or size—works effectively with or deliberately tries to include people of color, the rural poor, the politically and economically disenfranchised. . . . The leadership of the environmental movement stands as an obdurate white-male island in the middle of the work force increasingly populated by women and people of color. This is a most peculiar condition for a movement that is so firmly rooted in the tradition of American social change.[3]

Minorities are equally absent from the ranks of environmental and natural resource professionals in government, especially at the federal level, in agencies such as the U.S. Environmental Protection Agency and the Departments of Energy, Agriculture, and the Interior.[4]

Just as people of color until recently have not had a major presence in any part of the environmental movement, explicit reference to race, ethnicity, class, or to issues concerning the poor simply has not appeared in modern natural resource and environmental agendas. Conversely, and equally curious, until the last few years, minority organizations have scarcely protested these apparent omissions. There have been no marches, no media attacks, no boycotts. Just as race had not made the environmental agenda, it appears that environmental issues had not made the agendas of minorities and the poor.

When confronted with such information, one naturally tends to blame someone. How could large segments of our population not have a role in a sphere as important as the environment? How could environmental organizations, whose stated purpose is to improve environmental conditions for all, apparently neglect such a large segment of our society? How could civil rights and social action organizations, whose stated purpose is to improve the conditions of minorities and the poor, ignore such a large influence on those conditions? Here and in the next two chapters, I will address these questions and at the same time examine various dimensions of the above-stated conditions.

Who Are These People?

Let me first clarify the term *mainstream environmental movement*. While some critics from the environmental-justice movement tend to speak as if there were a single monolithic structure of mainstream environmental organizations with a shared agenda and beliefs, the truth is much more complex. Approaching these organizations as if they were one and the same is equivalent to assuming that all Republican organizations or labor organizations or civil rights organizations are one and the same. The range of beliefs, values, and goals within the many organizations that identify themselves as having an environmental agenda is very broad and often in conflict. In every aspect, these organizations vary widely—from their perception of the role of human society with respect to the environment, to their goals and methods of activism, to their definition of nature itself.[5]

Although wide variation exists among these environmental organizations, until a few years ago, almost all—whether mainstream environmentalists or "deep ecologists"—were either uninvolved or unconcerned with environmental justice or the particular environmental plight of the poor or minorities. Most did not even give lip service to the special environmental plight or burden of the poor and minorities. Some had not considered the issue and then rejected its validity, but rather had not even been aware that the issue existed.[6] While some of these organizations have begun to change, many still question the legitimacy of including in an environmental agenda issues such as unequal distribution of environmental hazards by race or income.

Furthermore, although there is a diversity of goals among environmental groups, they also share a common history. The most traditional national conservation groups, with over a hundred years of history, as well as ecology radicals of more recent origin, all trace their roots to the same "back-to-nature" bedrock of conservation of the nineteenth century. By virtue of this common history, these groups are here lumped together under the rubric of the environmental movement.[7]

For similar reasons, the term *environmental movement* also describes those nongovernmental organizations (NGOs) whose agenda, for better or worse, have helped define the environmental public-policy agenda as well as institutional and legislative responses to environmental issues. Until the emergence of the environmental-justice movement, the major national and international players in the creation and implementation of environmental policy have been the national mainstream environmental organizations, industry associations, Congress, and the corresponding agencies in the executive branch of government.

Half-Reasons and Half-Truths

Some would point out that these organizations' apparent lack of concern about race and class effects is not surprising, considering the conspicuous absence of people of color and the poor from the ranks of the mainstream environmental movement, especially at the professional and managerial levels.[8] For many critics, that underrepresentation within the movement begs for an explanation as much as the absence of concern for social justice or ethics does. Many of the same critics would maintain that the climate of these organizations does not make them attractive for nonwhites. Furthermore, if one condition leads to the other, then until the composition of the organizations' membership and leadership changes, changes in agenda should not be anticipated.[9]

In fact, such critics continue, the movement is almost as much to blame as the polluters themselves for the burdensome and threatening environmental conditions that minority and poor communities face. Because of a misplaced emphasis on things relevant to middle-class suburbanites, the environmental regulations and policies of this country are, or at least appear to be, skewed. Saving old-growth forests in the Pacific Northwest appears to attract more resources than lead paint poisoning and poorly maintained hazardous-waste sites in poor communities.

Of course, many others would dispute these accusations. They would counter that the environmental movement has a definite social consciousness, a concern for all groups. Aren't environmentalists, after all, more liberal and progressive in their politics and economics than the average citizen? Would socially conscious environmentalists create and maintain a movement that was harmful or even insensitive to the welfare of the socially and economically weak? Don't their environmental policies benefit everyone? The preserving of old-growth forest may promote the population's social and economic well-being as much as the eradication of lead paint.

But the fact remains that an explicitly stated appreciation for the unequal impacts of environmental conditions and policies has not been a part of the mainstream environmental movement in this country. How do these omissions correlate with these organizations' avowed high social consciousness? First, some studies, although not extensive, have investigated the relationship between an individual's environmental belief and his location on a left-to-right spectrum of political and social beliefs.[10] In general, these studies have concluded that while environmentalists lean slightly more to

34

the left than the population as a whole, overall political designations—such as left or right wing, liberal or conservative—have little to do with one's environmental attitudes.

On close inspection, there is nothing in the wide-ranging belief structure of the mainstream environmental movement that would, by its nature, compel a more compassionate or studied consideration of the welfare of particular groups of citizens. Contrary to environmentalists' claims to a heightened social consciousness, there is evidence among extreme environmental groups, such as the deep ecologists, of an underdeveloped concern for the welfare of pockets of humanity.[11] Steve Chase notes in his introduction to a small discussion series between two voices of environmental activism, Murray Bookchin and Dave Foreman: "there is clearly a misanthropic strain within the more extreme wilderness visions articulated by some deep ecologists. This blunts the social perspective and ethic of the entire movement and its members. Indeed, the deep ecology movement *as a whole* lacks a consistent or clear social analysis of the ecology crisis or even a consistent commitment to humane social ethics."[12]

And while the deep ecology movement is a small part of the environmental movement as a whole, the failure of the mainstream to repudiate or challenge that faction's extremely antisocial views taints the entire movement. But even at its center, the environmental movement in this country, in its modern history, does not have a strong social agenda or a significant concern with the social impact of environmental policies.

Who Is at Fault Here? More Half-Truths

Returning to the apparent exclusion of minorities and concerns about social impact from the process of environmental policy-making, some within the mainstream environmental movement would place the blame on minority and poor communities themselves, rather than on environmental organizations. These critics contend that minorities have never shown an interest in environmental causes and that the low representation of minorities in both the ranks and the leadership of the movement, as well as the absence of race or poverty issues on the environmental agenda, reflects that lack of interest. In other words, if a group is not present to articulate its needs and goals, then they do not get voiced. Perhaps issues such as civil rights, drug abuse, crime, or employment have a higher priority in minority and poor communities than do environmental concerns. It has also been hypothesized that because most minorities live in cities, they have never had an opportunity to appreciate nature and thus find it hard to support environmental causes.[13]

This explanation, of course, assigns responsibility to the victim and completely avoids recognizing any responsibility environmental organizations may have for these conditions or attitudes. This view also places environmental concerns in a very nonurban, back-to-nature context. And it makes the very telling assumption that race and minority concerns have no standing within the environmental community unless minorities and the poor themselves participate. But this position also carries within itself a poison pill. It is as if concern for the environment is a luxury that only the comfortable can afford. Curiously, acceptance of this position would also argue that because environmental issues have a lower priority among a large segment of the population than issues such as crime, education, and jobs, then environmental issues do not deserve as prominent a position in the national agenda as those issues.

Although this prioritization position may have some validity, it would be much too simplistic to attribute the low representation and participation in the environmental movement by minorities and the poor solely to their different prioritizing of issues. As Chapter 5 will explain, minority concerns about the environment in many ways are not so different from those of the majority population.[14]

Conversely, the explanation that the mainstream environmental movement is simply racist and too middle-class to include either members or concerns of minority or poor communities is also too simplistic. Yes, the racial and social insensitivity of organizations in the environmental movement may in part explain the admitted historical absence of environmental justice issues from their principles and goals. And yes, the largely middle- and upper-middle-class composition of the environmental movement's membership and leadership may in part explain their ignoring of the environmental effects on the poor, nonwhite, and politically powerless. But such a sweeping, simplistic indictment does not provide the whole answer.

Consider the reaction of poor and minority communities to mainstream environmental organizations. If racism, active hostility, or even indifference toward the disenfranchised were the only reasons for the current imbalance in the makeup and agenda of the environmental movement, one would have expected a much greater hue and cry over the years from those communities. But until the last few years, challenging the agenda or the power structure of mainstream environmental movements has not been a significant part—or in many cases, any part—of the agenda of national organizations devoted to social action or justice.[15] Of course, lack of attention from social action organizations is not the same as an endorsement, but it definitely is not a condemnation.

While perhaps no single reason can be identified, it is important to explore all the likely sources of the problem. Being too quick to condemn invites equally quick and simplistic solutions to the problem. Gaining a deeper understanding of why the current array of mainstream environmental organizations is so lacking in diversity may point to the role these organizations can be reasonably expected to play in the future of the environmental-justice movement. At the same time, such understanding may also indicate what role the growing environmental-justice movement may play in the future of the environmental movement as a whole.

They Came from the Woods

Failing to appreciate the history of the environmental movement could lead to some very misguided and ineffective policy decisions in environmental justice. But given that history, mainstream environmental organizations cannot reasonably be expected to either quickly embrace the environmental-justice doctrine, or to readily accommodate the organizational, ideological, or strategic changes that the concept of environmental justice necessitates.

Organizations, like people, can seldom, without great external force, be more than their history allows. Currently, national environmental organizations stand on a foundation of several layers of conservation activism in this country. The conservation movement itself, with its great opposing leaders John Muir and Gifford Pinchot, was in large part a response to the rapid industrialization of America in the late nineteenth century. During the last great period of the conservation movement, in the nineteenth century, the great American urban centers emerged and the primary economic power of the country changed from agriculture to industry—environmentally destructive industry. Muir and other conservation leaders of the time feared for the natural treasures of this country and organized to prevent their imminent loss.[16]

Entering the nineteenth century, most white Americans still saw natural resources—water, air, land, trees—as virtually inexhaustible. Water was perceived as neither a scarce resource nor one whose purity could be significantly threatened. Heavy production and extraction industries, especially steel production and mining, treated water and air as free commodities to be used without regard for the consequences to either other humans or the rest of the environment. Even the agricultural methods of the period were destructive and ignored what today we would call externality effects. For example, contrary to a popular myth we have inherited from pioneer days, the common method in eighteenth-century America of clearing land for farming was not to chop down trees one by one and then pull out the stumps. Usually, pioneers simply started a forest fire and then cleared the charred remains. The fire's effects on resident wildlife or on the environment outside the clearing were given little consideration. This method was not a European innovation but was copied from the Native Americans.[17] Throughout most of the history of the United States, whether from ignorance or indifference, the users of America's natural resources have ignored the consequences of that use for future generations.

When the movement did gain momentum, conservationists of the last third of the nineteenth century, mostly Eastern professionals, focused their attention almost exclusively on preservation, or what they termed the "wise use" of western lands.[18] The already highly cultivated land of the Midwest and East was not a major focus of their agenda. Then, as now, ranchers and farmers in the western states often accused outsiders of sticking their noses where they didn't belong.

Conservationists of the late nineteenth century deeply distrusted what they saw as the danger of indiscriminate waste and the corruption of American industry, especially the emerging giant monopolies of transportation, metal production, manufacturing, and mining. But another cornerstone of their ethos was a deliberate turning away from the urban life. They viewed the city as part of the environmental problem rather than part of the solution.[19] It was the squalor of the city and the demands that urban dwellers placed on resources that fueled the exploitation of the land. The conservation movement had no place in its nature-focused ethos for direct confrontation with urban life and no consideration for the problems of the poor, much less the problems of the people of color.

This antiurban bias did not begin with the late-nineteenth-century conservation movement. This movement, which gained a considerable following in the late nineteenth century, had firm roots in the Romanticism and Transcendentalism of the late eighteenth century and the entire nineteenth century.[20] In turn, Romanticism was itself a reaction to the Industrial Revolution in England, and later the United States. It rejected the complexity of urban industrial society for the spiritually more valuable natural world. The poverty and pollution of urban existence played a large part in the Romantics' rejection of urban life. The early conservation movement, with its Eastern base, drew upon both the earlier European Romantics and such American Romantics and Transcendentalists as Ralph Waldo Emerson and H. D. Thoreau. It is therefore not surprising that the early conservation movement sought to escape, not confront, all things urban.

Even today, most environmental and virtually all conservation organizations retain a basic distrust and disdain for urban life.[21] David Pepper notes that part of the modern ecocentrism was another page from the same basic script: escaping the city to the life of the country.[22] And it was from suburban America that the original recruits of the modern environmental movement came. A natural, although likely not a deliberate, consequence of the origins of the principal environmental supporters and the roots of their beliefs would be the exclusion of social justice issues from their agenda. In fact, given the antiurban and Romantic–Transcendental foundations of the conservation movement, it would be surprising had its offspring, the modern environmental movement, included strong social justice elements in its early agenda.

In most of Western Europe, the environmental movement evolved as it did in America, and it can also trace its roots directly back to the conservation movement. These two movements in turn greatly influenced both the underlying ideals and the character of nascent environmental movements in other parts of the world. Inherent in all these movements is a strand from the Romantic and Transcendental philosophies: the idealization of nature and the value of preserving nature in its undisturbed state. According to these philosophies, humans are often viewed as an intrusion in the natural world and their activities, especially the manufacturing and production of the post–Industrial Revolution of the late nineteenth century, as a despoliation of nature.[23]

Until very recently, the location of hazardous-waste sites or other environmentally risky activities has been considered by many environmental organizations to be an urban problem. Lack of concern over the inequity of hazardous-waste site location or similar issues of environmental justice was less a manifestation of racial or class bias than simply a rejection of all urban problems.

The Conservation Movement and the U.S. Public Health Movement

During the last third of the nineteenth century, when the conservation movement was in its infancy, urban living conditions were rapidly worsening. Urban pollution due to increased population density, indiscriminate industrial waste, and poor public sanitary systems made life miserable for many indigent city dwellers. These conditions provided the impetus for the public health movement in America and Western Europe.[24] Yet as noted earlier, the founders of the conservation movement, who were definitely both aware of and upset by these threats to urban dwellers, chose not to confront urban problems directly. Instead, they turned their backs on the cities and ran to the woods.

The conservation movement and the public health movement, both of which emerged in the nineteenth century, were two very different reactions to the extraordinary economic, social, and cultural changes then occurring in American society. Almost a century later, that difference has led to an environmental movement with little connection to the urban environment and a public health movement with only a tenuous connection to the world of natural resources.

From our early-twenty-first-century perspective, we can only speculate whether such a contrast was inevitable. Perhaps if the early conservation movement had not so strongly rejected urban life, or if the emerging public health movement had recognized the role of the natural environment in the well-being of humans, a very different environmental movement would have emerged after World War II. It is even possible that if the impetus for environmental reform after World War II had come from the public health movement instead of the conservation movement, the burden of specific communities would occupy a much larger place in the agenda of the modern environmental movement. Furthermore, of the two movements, the public health movement has always had a larger representation—both in its membership and its leadership—from among the poor and minorities. Thus, with a foundation in public health rather than conservation, the makeup of the modern environmental movement—both governmental and nongovernmental, policy-makers and professionals—would probably be much more representative today.

Even now, the division between the environmental movement and the field of public health rests on history rather than substance. In areas such as lead poisoning, indoor air quality, and municipal water treatment, the line between the two has become blurred. Aspects of such problems are claimed by both. Except for the most radical of the biocentric environmentalists, both environmental policy and public health policy today concern themselves with the interaction of humans with their environment.

Yet differences do abound. While governmental institutions are responsible for major environmental protection activities, environmental NGOs retain a significant role as information providers and activists. If asked, most people could identify at least two or three environmental NGOs, such as Greenpeace, the Sierra Club, the Wilderness Society, Friends of the Earth, or the Conservation Foundation. But how many could name even one nongovernmental public health organization? Public health has changed from citizen-led activism against unacceptable urban conditions to an almost exclusively governmental or institutional activity. Few individuals today join public health action organizations, yet many public health problems remain as pressing as ever. One possible explanation of the absence of large national public health networks or NGOs is that in its current form, public health administration is taken for granted by the middle class. Urban environments are no longer seen by the middle class as something to manage or remedy, but rather as something to escape by moving into the suburbs.[25] I speculate that it is possible that if the middle class had not abandoned cities for the suburbs after World War II, the character of both the public health and environmental movements would be very different today.

Speculation to the contrary, the fact remains that the two movements did not converge, and today, some factions of the environmental movement have moved even further away from concern over human conditions.[26] There is little evidence that anything but a traumatic external shock would reverse these attitudes. While some public health issues such as water and air quality do have a place on the agenda of many environmental organizations, most issues related to the human urban environment have few champions in the environmental movement.

The Future of Environmental Justice and the Environmental Movement

As evident in its historical foundation, the modern environmental movement contains within itself elements hostile to the healthy growth of a concern for social justice, such as responding to the unequal impact of environmental activities and policies. The force of this heritage still appeals to a large element in our society that continues to yearn for a romantic ideal of nature and wilderness.

As one examines the interrelationship (or lack thereof) between the environmental movement (as it existed into the 1990s) and issues of environmental justice, several patterns appear. First, the attitude and orientation of some mainstream environmental organizations have been definitely changing over the last few years. For example, in the 1990s, several mainstream organizations—notably the Sierra Club, the Audubon Society, Friends of the Earth, and Greenpeace—began to recruit minorities, both for membership and for staff and decision-making positions. Furthermore, the Sierra Club and Greenpeace, to cite only two examples, have also participated in aspects of the environmental-justice struggle by filing legal briefs in specific cases or by providing organizing or information resources. But do these actions signal the embrace of the environmental-justice cause by mainstream environmental organizations? And is such an embrace even welcomed by existing grassroots environmental-justice organizations? It is too early in this history to do more than speculate.

If such trends continue, can mainstream national environmental NGOs ever assume a leadership or major policy role in the environmental-justice movement? Recall that the environmental-justice movement started as primarily a local grassroots movement. Although some small national organizations have begun to form, the movement continues to depend on a slowly evolving network of loosely connected local groups.[27] These groups still see their plight as local and have no strong sense of a national or institutional agenda. Of the more than four hundred organizations listed by Robert Bullard in his environmental-justice directory, the vast majority are locally based and oriented. And the links between these organizations and the much larger environmental organizations in the national mainstream remain tenuous.

Given the historical antiurban ethos and the attitude of the current membership of most national environmental organizations, we should not expect organizational, political, or intellectual leadership in environmental justice from these groups. While they have the potential to provide some resource support, they will probably never completely throw off the ideological constraints that define them.

On the other hand, environmental-justice issues will no longer remain the concern of only minorities, the poor, or the politically disenfranchised. Very likely, they will become part of the policy agenda of many mainstream environmental organizations. But because these groups are burdened with understanding, supporting, or championing the cause of environmental justice, the intellectual impetus and policy leadership for those issues will much more likely come from outside these organizations.

This change in the underlying paradigms of the environmental policy agendas also means something else. As noted elsewhere, mainstream environmentalism in this country has been dominated by a white, middle-class, educated membership. But suppose that the primary environmental issues in this country and elsewhere change, as predicted. Then not only must the movement change its agenda to accommodate, or at least react to, the demands of these issues, but it must also change the composition of its membership. Current mainstream NGOs must greatly diversify their membership and leadership if they hope to retain anything like their current prominence. While it is not possible to accurately predict the outcome of such a change in environmental issues, these groups could retain their large memberships but become increasingly irrelevant as voices of policy. Certainly, many in the current environmental movement would resist to the end any attempt to dilute their back-to-nature orientation. And some of these groups and individuals certainly have the resources to resist such change. However, their opposition may very likely become marginalized.

Furthermore, minorities will probably increase their participation in mainstream environmental organizations. This increased participation, however, should not be confused with solving the environmental-justice problem. That solution will center on activities and organization outside the realm of these mainstream environmental groups. The intellectual roots and political foundation of the minority and poor groups that form and will form the backbone of this movement are different from mainstream environmentalism and will probably remain so for quite a while. This difference is neither good nor bad; it simply is.

As they evolve and expand both organizationally and financially, the grassroots network of environmental-justice organizations—with their currently strong representation among minorities and the poor—could rival or supersede these national mainstream environmental groups as the primary voice of environmental reason in this country. The most likely scenario is that the grassroots environmental-justice organizations, which face their own set of organizational well-being critical points, will grow at the national level and become organizationally more sophisticated as well as more professional in the execution of their agenda. According to the same scenario, the groups currently in the environmental mainstream will change dramatically: they will become more sensitive to the social impact of environmental policies and activities and more diverse in their membership and leadership.

To successfully confront the next generation of environmental questions, those organizations now in the mainstream of the environmental movement must deal with issues of environmental justice at all levels, both national and international. Within a very short time, mainstream organizations must recognize that environmental justice is an intrinsic part of almost all environmental decisions and actions.

Four

The Evolution of Environmental Justice as a Policy Issue

A Movement Whose Time Has Come

Issues of environmental justice were not on the agenda of the original environmental movement. Yet by the late 1990s, environmental justice had become one of the emerging themes of environmental policy discussion, and the topic promises to become a dominant theme of environmental policy in the early twenty-first century. These issues involve agencies, government activities, and policy issues not traditionally considered part of the environmental debate or agenda. But some of the problems encountered today in environmental justice stem from an inability or unwillingness by many in the mainstream environmental movement and elsewhere to evaluate environmental policies and action in terms of their unequal effects on the health, economy, and social structure of different populations.

Concerns over the differentiated impact of environmental activities and agendas were conspicuously absent from environmental policy in the 1960s and 1970s.[1] Even by the end of the 1970s and the beginning of the 1980s, the environmental-justice movement was little more than a few committed community organizations and individuals acting alone, with little national recognition. At the beginning of the 1980s, few in either the general public or the environmental mainstream had even heard of the topic, much less felt a need to address the problems that accompanied it. Even by the end of the 1980s, many in the mainstream environmental movement openly questioned both the legitimacy of environmental justice as a problem and, beyond that, questioned their responsibility for participating in its resolu-

43

tion. Yet by the beginning of the 1990s, specific legislation, agency policy, regulatory rule-making, and judicial determination in environmental justice were actively being discussed at the federal and state levels. In a few cases, specific remedies had already been enacted or regulated by the end of the 1990s.

So what propelled environmental justice from relative obscurity to an advanced stage of policy development where legislative formulation is actively and realistically discussed? The mainstream environmental movement had ignored environmental justice (or environmental racism, as some call it) for more than thirty years; why has it become such a hot topic now?

Evolution of a Public-Policy Issue

To answer these questions, we must first consider how any public-policy or social issue is formulated.[2] How and why does a particular condition become a matter of public-policy debate and action—particularly national public policy? Why do some public problems become public issues and not others?[3] Why do some problems, although often discussed by the public, never develop into a specific set of public-policy agenda items, or if they do, why are they extremely limited in scope? Why do we have a national policy on hazardous-waste disposal but not regional policies on managing truancy in the public schools? Why was there vigorous debate, including violent demonstrations, over a national military draft policy in the 1960s and 1970s but virtually no mention of the topic today?

Social scientists, particularly political scientists, have proposed a wide range of competing theories to explain the formulation of public policy. Some depict it as the outcome of struggles among the interests of various groups,[4] while others see it as the systematic expression of the values of a governing elite[5] or as a political system's response to a variety of situational activities and demands, both public and private.[6] In the same way, the manner in which policy decisions actually get made after the policy-formulation process has been described by some as largely incremental, with occasional disjointed departures[7]; a contrasting interpretation describes the actual decision-making process as an evolutionary selection process in which an issue finally emerges from a "policy primeval soup" of ideas, both good and bad.[8] At its ideal, it has been described as a rational-comprehensive process in which largely measurable criteria are applied to select the outcome that maximizes policy objectives[9] or as a mixed-scanning process in which both rational-comprehensive and incremental strategies operate.[10]

A common thread in all the descriptions of the formulation of public policy is that it proceeds through a number of stages. These stages have

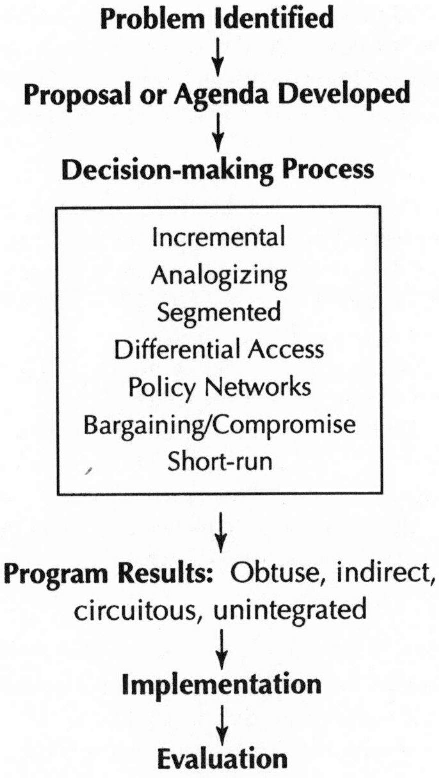

Problem Identified
↓
Proposal or Agenda Developed
↓
Decision-making Process

Incremental
Analogizing
Segmented
Differential Access
Policy Networks
Bargaining/Compromise
Short-run

↓

Program Results: Obtuse, indirect,
circuitous, unintegrated
↓
Implementation
↓
Evaluation

Figure 4.1. Policy development process.

been described in many ways by different authors,[11] but most agree that regardless of assigned labels, recognition of the problem must precede all other stages. Next, an agenda must be formulated around a particular set of responses to the problem. Then some subset of that agenda is formalized, and the subsequent policy, usually quite changed from the original agenda, is implemented. In some cases, the implementation stage is followed by the evaluation of the policy. Charles Jones describes the process as the "American way of making policy," and Figure 4.1 illustrates his interpretation of a six-stage process, beginning with problem identification and ending with program evaluation.[12]

However, almost all who have studied the process of policy formulation, including Jones, agree that in the real world, the creation of public policy seldom follows such a neat linear progression. For example, occasionally a policy is formalized even when a specific agenda has not been formulated. This was the case with the federal government's response to organized international terrorism. Both public and private legislation condemns such activities, but little was initially known about how to address the problem specifically.

Furthermore, not all policy or social issues jump through all the hoops. For example, when an issue results from the implementation of earlier policy, its formulation often includes only the implementation and evaluation stages. Conversely, many policy issues go through the program-recognition and agenda-formulation stages and then stall. Still other issues may skip over stages or occasionally backtrack.

Nevertheless, for the purposes of the present study, I will assume that an issue that evolves to the point of generating a public-policy response has gone through an identifiable series of developmental stages. For different issues, these stages will vary in length and even sequence. There may be stillbirths, retardation at the developmental stage,[13] and even an unexpected demise at the apparent peak of health. However, the nature of the stages themselves remains essentially unchanged.

How It Really Starts

To begin, even before the problem-recognition stage, something must occur—a change in economics or other societal forces—that turns an already existing condition into a problem of sufficient magnitude to merit policy attention. This merely implies that before such a change, the condition had not gained the status of a public issue. For example, many rural farming communities have little problem with the occasional presence of the smell of manure. However, as city growth in an area extends further out into the surrounding countryside, attitudes change. The smells associated with a working farm can then pose an environmental quality problem: a problem that now may require a policy decision and a regulatory solution.

So who cares? Why does understanding the life cycle of a policy issue have any importance except to academics and historians? From the perspective of policy analysis and implementation, the answer is clear. Appreciating when, how, and why an issue moves to another stage of policy development can actually help participants guide that process. Recognizing which stage (or stages) an issue currently occupies, and which stages it has already moved through, permits both its supporters and detractors to apply themselves only to those activities appropriate for that stage.[14] For example, ten years ago was not the time to expend energy on appeals for specific legislative remedies for environmental justice. Neither the public nor policy-makers were remotely ready. Similarly, today's detractors, who still challenge the legitimacy of the environmental-justice movement, very likely are wasting both their time and resources. The opportunity to remove environmental justice quickly from public attention has passed.

The environmental-justice movement finds itself at the crucial period at which a specific agenda for its key issues must be formulated. But as a policy becomes more sharply defined—so that a specific policy agenda emerges (for better or for worse)—the door closes on many alternatives.[15] Seldom can this reduction of choices be avoided, but an appreciation of the dynamics of the process can guide its evolution. Closing doors to alternatives too early can mean the loss of a powerful choice in the future. At the same time, maintaining some choices past their point of viability wastes organizational energies.

Obviously, issues of social and economic policy evolve through many stages of public awareness and policy response. In some cases, an issue and its associated policy agenda never move past one of the early stages. The evolution of complex issues, however, gives rise to the regeneration, modification, and synthesis of one or more of the original issues. The environmental-justice movement, for example, in part represents a synthesis of elements of the early modern environmental movement and the civil rights movement. And the modern environmental movement itself has its primary roots in the late-nineteenth-century movement for the conservation of natural resources in this country and parts of Europe. That conservation movement can in turn be traced back to the Romantic movement in Europe in the late eighteenth and early nineteenth centuries.[16]

Although the evolution of policy issues is complex and dynamic, patterns can nevertheless be loosely mapped. Of course, all issues, including those of the environmental-justice movement, do not follow the same pattern. But the recognition of their patterns does provide us with powerful insights both for possible future directions and for those developments that are least likely to occur.

Seven Stages of Policy Evolution

Adopting a convention that may be somewhat arbitrary, many researchers have described policy as moving through seven stages:[17]

1. Structural changes create the problem or move it to a higher level of concern.
2. The problem is recognized as a separable issue or set of issues.[18]
3. Organizations are established around the problem issue or issues.

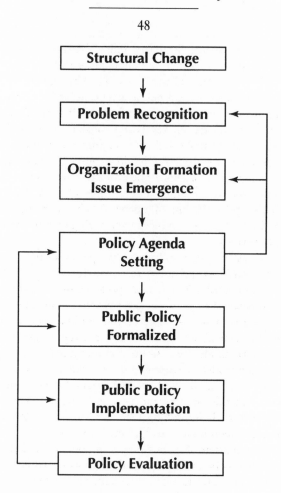

Figure 4.2. Life cycle of a social issue.

4. Policy agenda is set to deal with the issue or issues.
5. Public policy is formalized in a set of laws, regulations, or rulings.
6. Public administration of the policy response is implemented and becomes routine.
7. The program is evaluated.

Figure 4.2 describes the introduction of feedback loops between many of the stages—a modification of the linear nature of the description of the policy process. The feedback loops represent the outcomes often encountered when changing events in a later stage of policy development force a return to an earlier stage.

Stage One: Structural Changes

First, no social issue merely appears from the cosmic ether. Structural change—typically, a series of interrelated socioeconomic changes—must initiate the emergence of a social issue. This structural change can be perceived as upsetting a previous equilibrium. Others argue, however, that a state of equilibrium never exists; instead, there is a state of constant change and reinvention, and a particular structural change represents just one more turn of the wheel.[19] Regardless of interpretation, something must trigger the emergence of the issue. An important characteristic of this first stage is that the triggering change may be neither immediately recognized, nor may its connection with the social issue be appreciated.

Of course, eventual identification of the socioeconomic structural change that provided the impetus for an issue can play an important role in the later formation of a policy agenda. This identification can shed needed light on the potential scope and inherent limitations of any policy response to an issue. For example, a major force limiting attempts to correct the chronic unemployment faced by many of the urban underclass is that the traditional avenues for unskilled and semiskilled employment have largely vanished from the American economy. The steel, automobile, and other manufacturing-sector industries have changed the way they produce their product and how they do business. They no longer employ large numbers of semiskilled labor relative to their size of operation. Thus, even in prosperous economic times, these industries will no longer have a need for new and unskilled laborers. Appreciation of the structural changes contributing to the unemployment problem means then that the old policy response of simply pumping up the general level of national or regional economic activity will not address the problem of chronic underclass unemployment. Any new policy response in this area must then accommodate these limitations and constraints.

Another dimension of the life cycle of a policy issue is the level of public attention a problem enjoys at any given stage of its development. Some researchers believe that a social issue moves from almost no public attention through higher levels of public awareness, until it finally settles into a steady state.[20] Anthony Downs goes further, arguing that many public problems follow an "issue attention cycle." This cycle begins with (1) a "pre-problem state," before the public becomes aware that a problem exists; (2) a period of "alarmed discovery," with rapid increases in public awareness and a general consensus to take action; (3) a period, often involving intense debate, during which the "realization of the cost of significant progress" forces a reexamination of policy solutions; (4) a "gradual decline of intense public interest"; and (5) a "post-problem state," during which the level of public attention has settled into a prolonged limbo—a twilight realm of lesser attention or spasmodic recurrences of interest.[21]

After World War II, fundamental changes in Western (and particularly American) society, many of which are still not completely identified or understood, laid the foundation for the modern environmental movement. Americans had not suddenly become more moral or responsible than earlier generations; rather, they faced societal conditions that gave significance to environmental quality. The extraordinary increase in per capita real income during the 1950s and 1960s made environmental quality a high priority for an increasing segment of the population.[22]

Most Americans were now affluent enough that the quality of their immediate environment assumed a significant role in an individual's overall quality of life. As many commentators have noted, when disease, poverty, and war have reduced one's life expectancy to only a few decades, one does not worry about cancer-causing environmental hazards that would not likely prove fatal until well after the expected time of one's death. When one is worried about where the next meal will come from, one does not worry very much about the loss of an endangered species other than oneself. For most citizens in post–World War II America, thoughts of simple survival had been put to rest. The Great Depression of the 1930s was a fading memory for many. Now that tomorrow's meal was virtually certain, developing cancer (possibly caused by deteriorating environmental quality) in a decade or even two became a real concern.

This new affluence also resulted in a startling increase in the use of resources, which in turn led to a dramatic increase in the problems associated with such a lifestyle. As the economy grew, problems of waste disposal and pollution grew at a prodigious rate. Environmentalist Barry Commoner notes, "That period saw a sharp increase in the per capita production of pollutants. For example, between 1946 and 1966 total utilization of fertilizer increased about 700 percent, electric power nearly 400 percent, and pesticides more than 500 percent. In that period the U.S. population increased by only 43 percent."[23] At the same time, the amount of man-made materials used in daily life was increasing. During the period of around 40 percent growth in population from 1940 to 1960, the production of synthetic organic chemicals increased thirteenfold. During the following two decades (1960–1980), production of such man-made materials increased by a factor of 4.5, for a total increase from 1940 to 1980 of nearly sixtyfold, or 6,000 percent, while the population did not increase during the same period by even 100 percent.[24] As a by-product of this increase in the production and consumption of man-made materials, the production of environmentally threatening substances grew at an even faster rate than overall resource use.[25]

Furthermore, this period saw a massive population shift from inner cit-

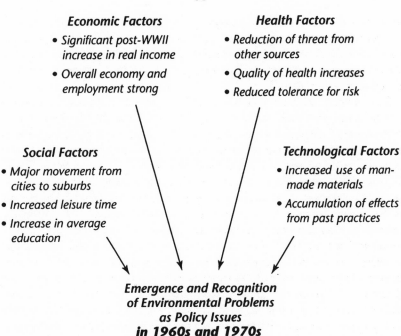

Economic Factors
- *Significant post-WWII increase in real income*
- *Overall economy and employment strong*

Health Factors
- *Reduction of threat from other sources*
- *Quality of health increases*
- *Reduced tolerance for risk*

Social Factors
- *Major movement from cities to suburbs*
- *Increased leisure time*
- *Increase in average education*

Technological Factors
- *Increased use of man-made materials*
- *Accumulation of effects from past practices*

**Emergence and Recognition
of Environmental Problems
as Policy Issues
in 1960s and 1970s**

Figure 4.3. Factors influencing the emergence of the environmental movement.

ies to the suburbs. The suburbs were ideally suited for this latest environmental movement with its foundation in middle-class needs and wants. The suburban lifestyle—with its increased space, a closer connection to nature, and much greater resource consumption per household—represented the modern ideal for many.[26]

Of course, affluence alone is not enough to explain the increased public concern with environmental quality. Many countries with a per capita income even greater than that of the United States have been infertile ground for the environmental movement. The key factor is the demographic distribution of that income. For example, in 1990, the per capita income in the United Arab Emirates was $19,870, whereas that in the United States was $21,810.[27] In America, a large segment of the population enjoyed the benefits of this economic wealth, whereas in the Emirates, much of the wealth remained in the hands of a very few. When wealth is widely distributed, political and social institutions permit the evolution of an issue like environmental quality.

Figure 4.3 illustrates some of the major factors influencing the rise of the modern environmental movement. Interrelated social, economic, technological, and health factors all contributed to an increased concern about environmental quality in this country in the 1960s and 1970s. And although actual environmental risk may not have changed, the tolerance of many individuals for such risk declined. This result in part reflects the effects of increased personal income and higher educational levels on population preferences.[28]

Furthermore, the late 1960s and early 1970s was a time of general social, political, and economic unrest in the United States. The Cold War of the 1950s and early 1960s was changing from a threat of nuclear annihilation to the local fighting of foreign intervention in Vietnam. Protection from nuclear war requires the strong presence of a central government. With a reduction in the perceived threat of nuclear conflict came a reduction in the perceived importance of support for centralized authority. Just as Americans were achieving the highest overall standard of living in the country's history, changes in international politics, the civil rights movement, and antiwar radicalism were altering the ways citizens related to authority. Increasingly, people were no longer as ready to accept that government knew best, or that "what's good for General Motors is good for America." Added to this growing disenchantment with authority, the subsequent Arab oil embargo of 1973, although coming after the creation of the U.S. Environmental Protection Agency (EPA), highlighted the vulnerability of ordinary citizens to the politics of natural resources.

However, the 1970s also passed. The civil rights movement had waned by the late 1970s, the antiwar movement did not last forever, and the Arab oil embargo did end. But the loss of faith in institutions of authority did not vanish. Citizens felt vulnerable to many things they had never given much thought to before. It was no longer enough to trust the government and industry to take care of things. Environmental needs were perceived as requiring specific action, not hope and trust.

By using this same argument of structural change, the environmental movement, as it developed in the 1960s and 1970s, with its agenda of "green grass and Bambi," had little appeal to many minorities. Absent from that movement was serious debate over the possibility that subsets of the population bear disproportionate environmental burdens. Mainstream environmentalism had not moved very far from its emphasis on the preservation of natural resources and the glorification of its roots in nature. Those roots did not include concern over the problems of the urban poor; in fact, a major feature of those roots was a rejection of modern urban development.[29]

Immediately after World War II, while many of the factors influencing the rise of the modern environmental movement also impacted minority communities, many others did not. Minorities did not—in fact often could not, due to discriminatory housing practices—join whites in their flight from urban centers to the suburbs. Also, whereas all Americans enjoyed significant increases in real income during this period, the actual size of the increases for minorities fell short of those gained by the majority.[30] In the same way, the gains in education for minorities that did occur came nowhere near those of the majority community.[31]

As the environmental movement arose among the American white middle class, the many minority communities in post–World War II America turned their focus to civil rights. For African Americans, the American South was, for most of the twentieth century, a police state where they had few rights and even less of a future. Other parts of the country, although less overtly repressive, were still extremely hostile and stifling sociopolitical environments for persons of color. For those people, economic, social, political, and educational opportunities and achievements were obtained only at great cost and great effort. Significant involvement of minority communities in environmental issues awaited later structural changes. And given the absence of such changes in the early days of the environmental movement, it should come as no surprise that the minority community scarcely was interested in mainstream environmental issues from the 1960s through the 1980s. When an individual is still worried about being arbitrarily arrested, being denied the basic right to vote, or seeking employment without unfair and widespread barriers of discrimination, that individual does not worry as much about the message of Rachel Carson's *Silent Spring*.[32]

But by the beginning of the 1980s, minority communities had realized significant progress in civil rights as well as economic and political opportunity. Some may argue that on several fronts, the decade of the 1980s was a period of stagnation, or even loss, with regard to many civil rights and opportunities. A federal administration perceived as unsympathetic to the aspirations of the poor and nonwhite fostered a social climate of intolerance and hostility. In spite of these negative circumstances, both real and perceived, many aspects of life had drastically improved, as measured by almost any yardstick, for significant portions of the minority population compared with what they had been thirty years earlier. This, of course, does not deny the crisis caused in most inner-city communities by the potential loss, both literally and figuratively, of an entire generation of youth, especially young men.

Nevertheless, structural changes in the socioeconomic landscape occurred that by the 1980s permitted or forced increased minority involvement in environmental issues (Figure 4.4). Changes in America's economic, health, and social structure—such as progress in civil rights and a concurrent deterioration of some environmental conditions directly affecting minority communities—revised the social equation. As a result, environmental is-

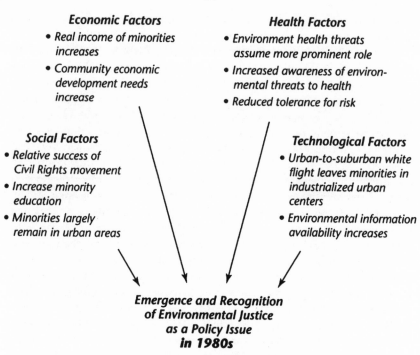

Economic Factors
- *Real income of minorities increases*
- *Community economic development needs increase*

Health Factors
- *Environment health threats assume more prominent role*
- *Increased awareness of environmental threats to health*
- *Reduced tolerance for risk*

Social Factors
- *Relative success of Civil Rights movement*
- *Increase minority education*
- *Minorities largely remain in urban areas*

Technological Factors
- *Urban-to-suburban white flight leaves minorities in industrialized urban centers*
- *Environmental information availability increases*

Emergence and Recognition of Environmental Justice as a Policy Issue in 1980s

Figure 4.4. Factors influencing the emergence of
the environmental justice movement.

sues assumed a far greater significance for the minority population, both for the daily life of the community and for the community's perception of its future well-being.

Just as the original environmental movement of the 1960s and 1970s sprang from the social and economic changes of the late 1940s and 1950s, the environmental-justice movement of the 1980s had its roots in the economic and political changes of the 1960s and 1970s. The contributions of these latter structural changes are not fully understood even today, but they would certainly include the following:

- The achievement of sufficient progress in other areas of social progress, such as basic political civil rights, gave greater opportunity for environmental-justice issues to gain minority attention.
- Much of the environmental-justice movement has focused on exposure to and disposal and treatment of hazardous material near minority communities. This focus is more than just an example of preexisting environmental concerns moving up the minority community's preference

ladder. It also reflects the tremendous increase in the generation, and associated need for disposing of, hazardous materials in this country after World War II. This problem has significantly worsened over the last three decades, making it more visible for everyone.

- During this same period, information on the significance of various environmental threats has become more sophisticated and accessible, helping to heighten awareness of the unequal distribution of environmental burdens. At the beginning of the 1960s, for example, the devastating effect of lead poisoning on humans, especially infants and young children, was not appreciated. But by the 1970s, there was little doubt that lead poisoning was a serious threat to humans, one that would affect poor urban dwellers particularly harshly. And unlike many environmental effects that have a long incubation period, the damage of lead poisoning to young victims is all too apparent within a few years of their exposure. Thus, once aware of its symptoms, minority communities could quickly observe its effects and seek appropriate remedies.

- After World War II, a profound change occurred in American residential patterns. Whites were abandoning inner cities, or core urban areas, for the suburbs. The expansion of American roads and highways assisted in this exodus by permitting a significant reduction in the home-to-work travel costs. With improved road systems, people could escape further from the inner city at a lower cost per mile, making living farther and farther away from their site of employment in the city feasible.[33] Through the flight to the suburbs, environmental threats, which previously had a more equal distribution among the many groups within the urban core, now affected the poor and minority urban communities that were left behind more heavily and more visibly.

  Interestingly, while many manufacturers ultimately also followed the population into the suburbs, industries—especially more traditional manufacturing industries—remained in the cities. Minority and poor populations began to perceive that it was their communities, and not the suburbs, where industry and its accompanying pollution were located.

  This developing disparity between majority and minority environmental experiences does not argue that there was any "deliberate plan" to saddle minority communities with the burden of waste and increased environmental risk. But the effect, nevertheless, was a steady increase in the inequity of environmental burden faced by those left behind in the urban centers. Urban flight had as one of its unexpected consequences an involuntary subsidy of the white and prosperous by the nonwhite and poor.[34]

- Like the larger majority community, the poorer minority communities did enjoy real per capita income gains during the post–World War II period, although those gains came years later.[35] While major economic, social, and political problems still handicapped these communities, they, like the white majority communities before them, may have become prosperous enough that they could worry about the level of environmental quality reached. Some evidence for this is found in the results from several surveys performed in the early 1990s on the importance of various "national issues." In these surveys, the percentages of whites and blacks identifying "environmental issues" as "important" were very similar.[36]

At the same time, the overall level of educational attainment in minority communities had been increasing since the end of World War II. Higher educational attainment appears to go hand in hand with increasing environmental quality sensitivity. Increasingly, minorities were in a position to appreciate the seriousness of the environmental threats their communities faced.

- Except for lead and other highly toxic substances, many serious environmental hazards are gradual in their impact. Furthermore, while the causes of many such hazards were developing in the 1940s and 1950s, their effects in many cases would not be appreciated until the 1970s and 1980s.

- A more subtle change affecting many aspects of American society was the increased availability and distribution of information. Many communities were unaware of waste disposal sites and dangerous chemical production operations in their midst. It is hard to become concerned over a problem that one is not even aware of. However, by the 1970s and 1980s, fewer communities could be duped into mistaking a hazardous-waste disposal site for something innocuous.

- In another curious spin-off from other social issues, the civil rights movement gave a boost to the emergence of environmental quality as an issue by providing an organizational blueprint for how to bring a community together to voice protest, to put together a policy agenda, and to effect change. Much of the present environmental-justice movement takes its organizational form and approach from the early civil rights movement. Of course, given the other structural changes that placed an uneven environmental burden on minority communities, a movement would have eventually emerged in some form. However, the prior existence of a reasonably successful political action model allows for the much more rapid formation of a response, and a faster development to the second and third stages of the evolution of a social issue: problem recognition and establishment of organization.

The points noted above are definitely not the only structural changes that fostered the environmental-justice movement. Unfortunately, the genesis of the movement is too recent for any but the clairvoyant to properly identify all the major social and economic changes that led to its emergence. The more we discover about the structural foundation of the movement, the more we understand its logical path and its optimal agenda. Failing to appreciate the underlying structural changes that give birth to an issue (or set of issues) can often result in subsequent failure or inefficiencies in the later stages of its life cycle: agenda development and solutions. For example, when the early antipoverty program failed to recognize the structural changes that made the problems of the poor in the 1960s and 1970s different from those of the early twentieth century, many of the remedies it suggested were doomed to failure.[37] Likewise, those who devised the agenda of the early environmental movement were blithely ignorant of how deeply issues of environmental quality were rooted in structural changes in the economy of this country and the world. Their ignorance of economic realities resulted in many environmental policies and regulations that were both economically and operationally ineffective.[38]

Stage Two: Recognition of the Problem

One sign that a set of problems is evolving into a set of issues is the recognition that an identifiable and separate set of problems have developed, or may be developing, around a set of circumstances. Although the extent of the problems or their consequences may not be fully appreciated at this stage, at least it has been recognized that something needs policy attention.

What is amazing is that often, very few people are needed to recognize the problem. Pioneers are often perceived as crying in the wilderness, but once they have spoken, others often gather around and public awareness increases. Upton Sinclair's campaign to bring social justice to the food industry at the beginning of this century,[39] and the more recent pioneering work of Ralph Nader[40] in consumer product safety, were turning points in problem recognition. This is not to say that no one had raised a voice before these crusaders. In most cases, the particular issue was part of a larger preexisting system. Often the key contribution of these pioneers was that structural changes had occurred in society that made the issue in question independently important or necessitated raising it to a higher level of concern.

In the early twentieth century, Upton Sinclair brought attention to food contamination with his crusading and writings, especially his 1905 novel *The Jungle,* which exposed corruption in the Chicago meatpacking industry. Before Sinclair, there had been other efforts to control food adulteration and contamination, but the issue had not reached the threshold of national attention. At the beginning of the twentieth century, America was still an agrarian society and most people were close to their source of food.[41]

In such an environment, although food adulteration and contamination were present, they affected a much smaller portion of the population. But as Americans moved in increasing numbers into the cities and no longer raised their own food, the individual consumer had less and less control over the quality of that food. The structural change of increased urbanization, among other factors, made what had been a relatively unimportant issue into a problem of national concern. This demographic change gave impetus to a movement that led to federal regulation of food processing and inspection, and to the eventual passage of the Food and Drug Act of 1906 to carry out that mission.[42] Of course, what further aided the emerging movement for reform in food processing was the tacit approval of the large meatpacking corporations. These corporations recognized that any regulation would generate significant increases in both labor and capital costs—costs that their much smaller competitors could not afford.

Similarly, before World War II, a relatively small number of citizens owned automobiles. Such structural changes as the shift of urban population to the suburbs and the explosion in the national highway system led to an unprecedented increase in automobile usage.[43] In the 1920s, concerns over the safety of, say, the Ford Model T would have meant little to the average citizen. Few households owned automobiles then,[44] and the condition of most roads did not permit sustained high-speed driving. All this changed in the 1950s and 1960s. What had affected the few now affected the many, and concern about automobile safety moved from being a curious side issue to a central problem for American society. Ralph Nader was not the first to sound an alarm over auto safety. But his cry came at a moment in social history when the country was ready to hear it.[45]

Minority communities faced environmental threats well before the 1970s and 1980s. Many minority communities did protest against the location of facilities they felt were detrimental to community well-being—usually to no avail. But by the beginning of the early 1980s, a number of structural changes in American society meant that issues of environmental equity were ready for increased public attention and action, particularly within minority communities. The number of environmental-justice pioneers was typically small but, as in the creation of many other movements, pivotal. Perhaps the first indication of the movement's beginnings was the 1982 protest by a community group in Warren County, North Carolina, against a state proposal to dump polychlorinated biphenyls (better known as PCBs) in the county, which had the state's highest percentage of African Americans. The protest resulted in the jailing of over five hundred people, including U.S. congressman Walter Fauntroy (D-D.C.). A direct result of the Warren County protest was the 1983 study by the General Accounting Office (GAO) on all EPA Region IV hazardous-waste sites, including the one in Warren County.[46] The GAO report found both racial and income bias in the location of the sites.

The next significant warning came from Robert Bullard's 1983 empirical study of racial patterns in site location within the city of Houston.[47] A final key voice was the massive national study performed in 1987 by the United Church of Christ (UCC) Commission for Racial Justice under Director Benjamin Chavis, evaluating hazardous-waste sites nationally, which was a direct follow-up of the 1983 GAO study. Both the Bullard and UCC studies, at the city and national levels, respectively, found evidence of racial imbalance in the distribution of hazardous facilities. These two studies raised the national awareness level on a type of environmental issue unknown at the time. A key consequence of the UCC study was the recognition that the environmental problem an individual community faced was not unique to one community; rather, many minority communities across the country faced similar problems. Most significantly, the studies served to shift the field of debate from individual local communities to the national stage.

What emerged from these early works was the identification of a list of environmental problems that many minorities communities shared:

- Minority and low-income communities and individuals have definitely been left out of the decision-making process for environmental policy.
- These same groups are at a definite disadvantage with respect to input into the environmental policy process. Compared with the other players in the arena of environmental policy and regulation, they lack equal access to key information.
- Unequal environmental burdens exist—although in some cases, the inequity may be driven more by economic forces than racial intent.
- Environmental research has not historically been directed toward answering questions of environmental justice and is still woefully inadequate at suggesting solutions.
- The previously closely held belief that minority and low-income communities were not concerned about issues of environment, if ever true, is definitely false today.

Much of the present activity of the environmental-justice movement deals with these problematic positions.

An important feature of the problem-recognition stage is that besides heralding the real beginning of public attention to an issue, the form of this recognition also provides the font from which the eventual structure of the policy agenda will spring and evolve. The process of problem recognition is not static, however. The ramifications of an issue are never well understood or articulated at this early stage. However, constraints on knowledge do not reduce the absolute importance of identifying the crucial aspects of an issue. This is most relevant for those elements that are not intuitively obvious. The price for half-efforts or ignorance is high: failure to recognize key aspects of a problem can lead to a flawed policy agenda and solution.

Immediately after World War II, the environmental movement recognized that the quality of many citizens' lives was now inexorably linked to the quality of the environment. It also recognized that a major source of the problem was industry's use of environmental resources such as air and water without regard or accountability for consequences. But the early environmental movement did not completely recognize the powerful economic and social forces driving much of the emerging environmental problem. The movement also failed to appreciate the inadequacies of their chosen method of enforcement and regulation—the command-and-control regulatory approach—to correct these problems. These mistakes hampered the effectiveness of the subsequent legislation, regulation, and attainment of the movement's goals.[48]

Likewise, we must ask the following: Which issues of the environmental-justice problem have been ignored or incorrectly dismissed? Do some aspects remain undetected? Which of them could limit the effectiveness of this movement? Fortunately, problem recognition is an ongoing, dynamic process, and because the environmental-justice agenda is not yet chiseled in legislative granite, the possibilities for improvement are good.

What may cause greater concern are those aspects of the problem that are less obvious, not yet recognized, or simply denied. And within such neglect or denial lie the seeds of disappointment or failure. For example, an attitude of "not in my backyard" in the location of hazardous sites often will not work as a short-run national solution. Much of the current dialogue in the environmental-justice movement concentrates on the means and justification for preventing the location of such sites in low-income and minority communities. But short of a sudden, miraculous elimination of all pollution, hazardous materials must go somewhere. It is both naive and potentially irresponsible to act as if, shielded by enough laws or regulations, any community need never worry about these dangers. If the movement declares prevention alone to be a sign of its success, it is destined for disappointment.

Less obvious is that while the facilities' location burden is a local one, the only viable solution may require a national framework. Exclusively regional or state solutions simply shift the burden to the point of weakest local resistance or political influence across a national landscape. The absence of a strong national regulatory framework could actually worsen the location distribution problem by concentrating the environmental burden on that subset of the low-income and minority community with the mis-fortune to live in the wrong state or region of the country.

In a final example, the exposure of low-income and minority communi-ties to environmental threats such as hazardous-waste facilities, Superfund sites, and the release of toxic substances from manufacturing operations occupies center stage in the environmental-justice movement. The chief federal agency involved with the movement and with this particular issue has been the EPA. Yet in terms of total impact and actual (versus per-ceived) risk to communities, perhaps equal attention should also be paid to other risky conditions. More attention needs to be focused on less well known threats, such as the effects of pesticides on Hispanic farm work-ers,[49] incorrectly specified food contamination levels that do not reflect the consumption patterns of minority communities,[50] and the aging condition of housing and the substandard materials used in much minority urban housing, with its concomitant risk of lead poisoning.

Stage Three: Formation of Organizations

Once a problem is recognized, and if the momentum of public awareness and concern continues, organizations emerge. These issue-oriented organi-zations provide the backbone for much of the subsequent stages in the evo-lution of a social issue. Of course, given the enormous range of subjects contained in any social issue, the organizations that spring up in reaction to the issue problem or problems will focus only on a subset of related issues. Nevertheless, whatever the range of issues covered, organizations must emerge. Whether informal or formal, whether national, regional, or local, an organization must be formed in order for the issue to move to the higher level of public awareness and political influence necessary for the creation and implementation of a policy agenda.

Although individuals play a key role in the problem-recognition stage, the formulation of institutional policy must almost always pass through the organization creation stage. An interesting corollary to that axiom is that the less politically powerful or important the organizers, the larger or more broadly based the necessary organizations must be to effect change or to resolve the problem. The civil rights movement has involved hundreds, if not thousands, of organizations nationwide, with millions of active partici-pants. In contrast, the original environmental movement involved far fewer groups and a much smaller active cadre. Among other reasons for these differences, the former movement dealt with issues affecting the poor and politically weak minorities. In contrast, the latter movement, in its original form, campaigned for issues affecting the middle class and the politically influential.

The organizational establishment of Stage Three provides the fertile field from which broad and often heated discussion arises. From the limited domain of the interest of an informed few, the issue becomes the cause of many committed and formally established groups. In the environmental-justice movement, the initial focus was on correcting immediate local problems of location, removal, or correction of hazardous facilities. Broad national concerns came later. These emerging local groups lack funds and information resources and are only now beginning to form networks outside their individual environments.

It is interesting that initially, many of the locally based organizations first involved in the environmental-justice movement had targeted environmental justice as only one of the many community-based issues to focus on.[51] Community service groups, community health groups, and community residential improvement groups embraced environmental-justice causes as part of larger missions. A sign of the environmental-justice movement's maturation is that recently community groups have emerged that are dedicated specifically and solely to pursuing environmental justice.[52]

That most of the environmental-justice organizations came from preexisting local organizations with broader social justice agendas in many ways may have helped the emerging environmental-justice movement. Piggybacking on the organizational structure of these preexisting groups, the environmental-justice agenda enjoyed rapid mobilization and publicity. At the same time, these organizations had already established credibility within local communities, which benefited the movement. Through such association, environmental justice enjoyed a respectability and status that a completely new organization formed exclusively on the environmental-justice issue would probably not have provided.

It has been argued that the environmental-justice movement emerged rapidly because these local civil rights organizations, facing a waning in impetus of the civil rights movement during the Reagan 1980s, needed a new invigorating issue. The degree to which this is true has not been established. There is no question that the civil rights movement, after the major achievements of the 1960s and 1970s, appeared to have lost momentum and direction in the 1980s. That this loss can be ascribed to Reagan-era policies is not clear. With major political rights gains achieved during the 1960s and 1970s, the next stage of the civil rights struggle involved far more complex and intractable problems, such as education, social structure, and a changing American economic landscape with few opportunities for the unskilled. These organizations, often formed in response to the earlier and better-defined civil rights objectives, may have had difficulty adapting to these less well-defined problems.

That the environmental-justice issue may have proved a welcome new strategic direction for these civil rights organizations to venture into cannot be denied. But environmental justice, like the other remaining problems facing minority communities, also lacks the well-defined parameters and specific solutions of the early civil rights struggle. Thus, it is difficult to ascribe these organizations' adoption of the environmental-justice issue exclusively to political convenience. Why add one more difficult task to an already heavy load? In fact, during the 1980s, as the environmental-justice movement grew, the civil rights organizations that joined the struggle were initially almost exclusively local. The large national civil rights organizations, such as the NAACP and the Urban League, were not at the forefront of the environmental-justice movement in the 1980s. Clearly there was no national organizational movement to embrace the environmental-justice issue as a reinvigorating device. Local groups, on the other hand, faced situations in which they may have come to appreciate that environmental protection issues were a key component of the local social, health, and economic challenges they had to address.

Whether as a strategy for organizational strengthening or as the result of increased appreciation of the role of environmental quality in community well-being, environmental-justice issues were clearly helped in their policy development by drawing on the influence of the preexisting local groups. At the same time, the more recent emergence of local groups oriented exclusively toward environmental issues speaks to the development of the environmental-justice issue to the point of requiring or allowing more exclusive focus.

Stage Four: Setting a Policy Agenda

By the mid- to late 1990s, the environmental-justice movement found itself at the beginning of Stage Four: setting a policy agenda. Although some may still debate the legitimacy of the environmental-justice issue, due in part to weak empirical analyses, the issue itself and its perception by those for and against it have moved well past the legitimization stage. Once an issue enters the organizational establishment activity of Stage Three, debate over the imperfection of data can no longer bar it from the public forum. Short of a massive shift in public consciousness and the complete and magically silent dissolution of hundreds of community organizations, the movement will persist. This is helped because a major activity of the movement, even during the setting of a policy agenda, remains rooted in the previous stage: the formation and building of organization.

Roger Cobb and Charles Elder divide the development of a policy agenda into a "systemic agenda" and an "institutional agenda."[53] The systemic agenda represents all issues that require government action. The authors term this agenda a "discussion agenda"; it covers a broad range of issues that invite public discourse. In contrast, the institutional agenda includes those issues deemed by public officials to merit policy action. This agenda represents the formalization of working policy. Furthermore, within institutional agendas, old items—whether they be national, state, regional, or local—tend to be favored over new items. This makes sense for a number of reasons. First, given the incremental nature of at least major parts of the policy-formulation process, old items, which have served their time, have moved up through the priority system. Policy-makers thus often presume that old items warrant more attention because they have more familiarity with those items.

Following Cobb and Elder's interpretation, environmental-justice policy initiatives face a major stumbling block to their rapid legislative or regulatory adoption, due both to the difficulty of getting specific agenda issues onto various decision-makers' institutional agendas and to the newness of the field of environmental justice.

An interesting contrast may be made between the agenda development and perception of the environmental-justice issue and the issue of global warming. It has been argued that as an issue, environmental justice has moved past the point where its legitimacy can be challenged on the grounds of fuzzy analysis or insufficient proof. But the issue of global warming, which slightly preceded environmental justice as a policy issue, has not advanced past such challenges. Yet far more analysis and many more government research dollars have been devoted to the study of global warming than to the identification of environmental justice. Entire institutes at various universities and nonprofit organizations have sprung up to study global warming. Innumerable government reports, articles in scholarly journals, and miles of paper in the popular press have focused on this environmental issue. On the other hand, few academic institutes presently study environmental justice exclusively,[54] the number of scholarly works in the environmental-justice area is quite limited, and popular press attention has begun to increase only within the last several years.

These differences reflect in part the nature of the two issues and in part the anticipated costs associated with them. From its inception, the study of global warming has received substantial support from the environmental movement, but support from the academic science community has been weak and at times has appeared nonexistent. Proper prediction of global warming effects requires computational power and sophistication beyond the capacity of current methodology.[55] Major questions still remain about whether empirical observations truly constitute a global warming trend. The validity of current models is still in question. Whether merited or not, the global warming movement has always carried the burden of having to repeatedly justify its existence.

In contrast, challenges to the legitimacy of the environmental-justice movement have tended to focus on whether differences in environmental threats were examples of racial bias or economic bias. Virtually no one has challenged the proposition that some segments of our society bear a greater environmental burden than others. Thus, although it lacks massive analytic support, the environmental-justice movement has in some ways surpassed global warming in the establishment and implementation of a policy agenda.

A second significant reason—and quite possibly the major reason—for differences between the agendas of the two sets of issues is the underlying cost of properly addressing the problems. Virtually all agree, if they take the global warming scenario seriously, that the solution will require a massive, perhaps historically unprecedented, economic and social expenditure. And the cost must be borne not just by this country but by all nations and peoples on this planet. To reverse or even slow this process would require fundamental and generally unpleasant changes in the lifestyles and organization of all developed nations. Citizens in each of those nations would face a world in which continued improvement in the quality of their lives would no longer be certain.

It is not surprising that while many may wring their hands over the prospect of global warming and the ensuing changes in world order, no nation or group of nations has sufficient motivation to embrace the costly and traumatic changes necessary to control it. Given the apparently enormous economic and social costs of reversing global climatic change, the level of necessary proof has been set very high for this issue. By the same token, if global warming is a legitimate problem, the future cost to global well-being is sufficiently high that extensive research efforts will be willingly supported. Both the support and the doubt will probably continue in their pas de deux for several years to come.

Things are simpler for the policy agenda of environmental justice. Environmental justice, although a universal problem, can be addressed at many levels, not just globally. National, regional, and even local actions and initiatives can successfully address major aspects of this issue. More important, the costs associated with environmental justice are not as well defined, nor do they even appear to occupy as large a portion of the current issue discussion as they do for the global warming discussion. The proper pursuit of an environmental-justice solution may involve some increases in the cost of production, but those costs are still not well defined. In fact, from the perspective of efficient markets, these corrective costs may well be justified. That is, they reflect the true cost of producing certain goods and services that heretofore have enjoyed an involuntary subsidy from the affected communities. And because current production and consumption behavior in these markets is resource- and price-inefficient, the higher cost would be offset by more efficient resource use and pricing in the economy as a whole.

It will be interesting to observe the reaction that occurs as environmental justice moves into a stage of implementation where the many costs associated with addressing this set of problems become more specific. It will also be interesting to see how those citizens whose environmental and economic well-being has effectively been subsidized by overburdened communities react to regulations and mandates that reach into their pocketbook. How much environmental justice will America decide it can afford?

At this stage in the life cycle of a policy issue, the question is no longer, "Is there a problem?" The question instead is, "What do we do about it?" At this level, a specific set of policy recommendations should be formulated. But the diverse organizations orbiting around any set of issues will seldom generate a single set of recommendations. Within the environmental-justice movement, recommendations cover the entire spectrum, from the elimination of all environmental hazards facing poor and minority communities, to a simple provision that federal, state, and local government agencies be more sensitive to the problems of these communities when making future decisions. At the same time, all recommendations for environmental justice must compete for public support with issues as diverse as gun control, abortion rights, sexual harassment remedies, and health care reform.

The environmental-justice movement is also handicapped because its issues affect politically disenfranchised groups. The perennial concern over crime in the streets tends to be highly correlated with the number of white middle-class citizens affected. This bias is inherent in all aspects of the public-policy process, from the degree of serious consideration given a set of policy issues in the public forum to expectations about the level of resource allocation such issues can realistically expect.[56] The environmental-justice movement can achieve, and to some extent already has achieved much in spite of this bias, but the challenge remains.

Another reality of the policy agenda formulation process, which the environmental-justice movement cannot escape, is that the range of serious policy alternatives is constrained in several ways. First, except in unusual circumstances, policy alternatives must fall within the currently acceptable parameters of political and social behavior.[57] For example, an economically rational remedy for a community harmed by a polluting facility might be for each affected household to receive a direct payment from the offending source, with the amount based on the specific level of harm that household experiences. A less rational remedy, but one clearly satisfying the uniform distribution of risk criteria, would be to randomly assign such facilities to communities. The affected communities could receive no compensation and would not have any recourse to oppose the siting. Of course, neither of these policies would be seriously considered because both are well outside current social, economic, and political norms.

This range of "acceptable" alternatives is flexible and dynamic. Societal changes beyond the range of the behavior in question can produce changes in what is acceptable behavior and what is not.[58] But movement at the fringes seldom comes without effort. For decades, economists have argued that a direct cash payment (such as a negative income tax) to low-income individuals without the bureaucratic inefficiency of a food stamp or housing subsidy program makes the most economic sense and would be the best way to improve the welfare of low-income populations. However, a national program of negative income tax has never had the necessary political constituency and probably will never be viewed as a serious alternative for social assistance. On the other hand, the idea of private firms replacing public service providers in areas previously off-limits to the private sector—such as public education, police protection, and prison maintenance—is no longer a marginal idea, and in fact has actually been implemented in several communities around the country.[59]

For the environmental-justice movement, the key challenges at the stage of policy agenda development are to continue to gain public attention and credibility. Competing policy agenda alternatives have and will circulate, but all alternatives that seek a serious hearing must share several characteristics. Oddly, the likelihood that a proposed agenda will solve a given problem is actually not a top priority; it is more important that it be feasible both politically and economically. For example, proposals that would bankrupt whole industries or large numbers of companies will probably not be considered. Proposals that step outside the country's concept of governance or political self-image have little chance of moving past the speculative discussion stage.[60] Proposals that would require great political courage on the part of lawmakers with politically weak constituencies also have little chance, as do proposals that go against strongly held cultural norms.

A policy agenda may violate some of these axioms only when the issue is so compelling and circumstances so restrictive that no other alternatives will do. Such an example is the short-term application of price controls by the Nixon administration during the early 1970s as a reaction to the Arab oil embargo and the concomitant high inflation. The environmental-justice movement, however, does not appear to fall within this exceptional category. The public does not perceive the issue as a crisis, and compelling arguments for extraordinary measures have not emerged. Unless radical change in circumstances occurs, the policy agenda for environmental justice will probably stay well within current political, economic, cultural, and social norms. This is neither bad nor good; it simply states the current condition of both the movement and the public's perception of the issue: environmental justice is not a top priority for any major public interest group.

Stage Five: Formalization of Public Policy

Assuming that public awareness continues to increase and that the issue remains a valid one, a subset of the competing items of a policy agenda will be formalized as law or regulation. At this stage, the public awareness of the problem is usually at its highest and most vocal. Concern and support for redress of centuries of discrimination was at its highest in the mid-1960s, when the original Voting Rights and Civil Rights bills were passed. The Gramm-Rudman-Hollings Budget Reform was passed in the 1980s, when much of the public perceived the federal government to be in a budget deficit crisis. In the early 1990s, the push for universal health coverage in America emerged in a climate of mounting, almost unanimous concern over its availability and cost.

From the perspective of the 1990s, it is interesting to note that the subsequent major changes in the original Balanced Budget Act, which effectively removed what few teeth it had, occurred with little fanfare or public outcry. In the early 1990s, most of the public no longer felt that the budget demanded immediate and possibly radical action. But by the mid-1990s, the national budget deficit had once again moved to center stage as a new Republican majority in Congress dismantled or severely reduced many federal social action programs under the banner of deficit reduction.

In spite of a 1994 federal executive order, numerous bills in Congress, and several state efforts, the environmental-justice movement has not entered the stage of policy agenda formalization. Although it is on the threshold of policy formalization, parts of the movement remain in the policy agenda formulation stage, while yet other features of the movement are taken up with the still earlier stage of organization formalization. Statements of support for the concept of environmental justice have been offered, but very few specific remedies and virtually no funds for enforcement have been forthcoming. And, like the budget deficit in the early 1990s, neither the public nor lawmakers regard the issue of environmental justice as a crisis that necessitates radical remedies.

As a social issue moves into the policy formalization stage, several factors can create stagnation or even push the issue back to the previous stage of agenda formation. The issue's significance may fade due to changes in socioeconomic conditions, or the policy agenda may encounter a drop in support or even outright opposition. Federal civil rights laws had been offered decades before the passage of the 1964 Civil Rights Act, but active opposition and adherence to political norms about "states' rights" and other artifacts of our republican system prevented passage of significant legislation. The environmental-justice movement could face similar opposition from industries protesting a perceived attack on their right to engage in free enterprise or from an environmental establishment that denies the validity of the issues themselves.

Perhaps the most important achievement at this stage in the life cycle of a social issue is not just that a specific set of laws is passed or particular regulations promulgated. It is more important for the issue itself to have entered the national consciousness. Prominence on the national stage will change the way people view events. Although the Equal Rights Amendment did not pass and many legislative initiatives of the women's movement did not succeed, the issue of women's rights has fundamentally changed the way American society conducts itself. The same can happen to the environmental-justice movement. It is virtually certain that no specific legislation will be passed that will seriously address the vast collection of inequities under the umbrella of environmental justice. But as the issue of environmental justice moves into the stage of policy agenda formalization, it is even more improbable that the issue will completely fall out of the public consciousness.

Stage Six: The Policy Agenda Becomes Institutionalized

At some point, part of a serious issue's policy agenda actually may become law. Regulation may result. At that point, the immediate problem is no longer the focus of attention; rather, the focus becomes implementation. Having a specific set of proposals made into law does not guarantee resolution of the issue. In fact, because public attention tends to decline slightly at this point, groups may incorrectly perceive that their job is complete. As difficult as it is to raise public support so that an agenda is taken seriously enough to have institutional changes approved in the political arena, the task of implementing those steps can prove even more difficult. The Civil Rights Act was passed in 1964 and *Brown v. Board of Education* was decided in 1952, yet several decades later, implementation of these formal steps has suffered many disappointments. Numerous reasons have been posited for the failure of the racial justice legislation of the 1950s to remedy the discrimination and inequity rooted in American social and economic life. Chief among those reasons was a failure on the part of many in policy-

and decision-making positions to fully appreciate how entrenched racism was in American society.

Likewise, any environmental-justice legislation, should it pass, will likely not solve the problem. The failure of legislation to solve the problem often reflects a failure to understand all its social dimensions—aspects of social behavior well beyond the original understanding of even its strongest advocates. And the problem's solution will almost certainly require a battery of legislative and regulatory remedies well past any originally considered.

Institutional success also depends greatly on the support given to legislative mandates and regulatory remedies by bureaucratic agencies. In the early to middle 1960s, serious federal executive effort supported the introduction of the Program Planning and Budgeting System (PPBS).[61] After limited success in a few federal agencies, this new method of budgeting was imposed as the paradigm for most federal agencies. Unfortunately, many old-line federal agencies were not willing or ready to adopt what was perceived as a radical and potentially threatening method of budgeting.[62] Agencies throughout government placed obstacle after obstacle in the path of the system's adoption. In only a few years, the complete conversion of the federal government to the PPBS was abandoned.[63]

Given the history of the environmental movement in this country, there is every reason to expect that whatever remedial legislation is passed, some agencies will oppose its implementation. Sensitivity toward issues of social justice has no historical precedent in this country, and most of the federal agencies responsible for environmental protection do not operate in an organizational culture that places a premium on such issues. Despite lip service, there is little likelihood that rank-and-file environmental workers will quickly embrace such a radical departure from their organizational norms and ideals. Organizational resistance to environmental-justice initiatives should not be a surprise but rather an assumption.

Stage Seven: Program Evaluation

Of course, the development of environmental-justice policy is nowhere near the stage of general program evaluation. Few regulatory, and even fewer legislative, mandates exist, and those that do exist are too recent to merit serious discussion. The only environmental-justice program (if it can be called that) is the collection of agency actions taken after the Presidential Executive Order of February 11, 1994. By and large, agency response to the executive order has been to create system opportunities for introducing consideration of environmental-justice issues into agency policy decisions. It is still much too early to assess the significance of these opportunities.[64]

In the evolution and development of a policy issue, program evaluation and reexamination is a natural turn-around stage. That is, in an ideal world, after program evaluation, the policy approach and even goals may be modified. Realistically, program evaluation can take many forms, both formal and informal. The key feature is that in this final stage, questions about program effectiveness, mission accomplishment, and impact on society in general get raised. They may not always be answered, but they are raised.

Five

Misconceptions about Minority Attitudes toward Environmental Issues

Around here an environmental problem is not having your kid shot.
—East Harlem landlord, 1992

Minorities, especially African Americans and Latinos, are virtually nonexistent among both the leadership and the rank and file of America's mainstream environmental movement. It is as if environmental problems affect only whites—and middle-class whites at that—so only they have become involved in the movement. Alternatively, it seems that only whites have acknowledged the importance of the environment to society's well-being. Equally disturbing, as explained in the last two chapters, environmental-justice issues have enjoyed little prominence in the agenda of mainstream environmental organizations. Are these two conditions connected?

In this chapter, I will begin to explore reasons for the underrepresentation of minorities among the rank and file of the national environmental mainstream. This chapter and the next will show that minority membership in both governmental and nongovernmental environmental agencies and organizations, as well as input into their agendas, results from much more than supposed minority indifference to environmental issues.

Explanations

During a case study on environmental justice in the late 1980s, one graduate student at a major professional school for environmental policy and science responded that "Blacks and Latinos have never shown an interest in environmental affairs, it just doesn't matter as much to them." This response reflects a commonly held belief. Furthermore, many critics and supporters of the environmental movement believe that as a result of the lack of mi-

72

nority involvement in the movement, for whatever reason, environmental issues that exclusively affect minorities receive little attention. It is also argued that mainstream environmental organizations are closely connected to the biological and chemical sciences, and minorities, especially African Americans and Latinos, are poorly represented within those fields. It should therefore come as no surprise that minority participation is low in both governmental and nongovernmental environmental organizations.

A cursory inspection of the racial composition of national environmental organizations appears to support the above assertions. Minority representation in the policy and senior government service levels of such federal agencies as the U.S. Environmental Protection Agency (EPA) and the U.S. Departments of Interior, Agriculture, and Energy falls below the average for federal agencies.[1] A 1993 EPA survey on cultural diversity asked respondents why they had chosen to work for the EPA. Seventy-one percent of the white respondents chose the response, "To help protect the environment"—as opposed to only 42 percent of the African Americans.[2] The implication is that even those minorities who did work for government environmental agencies were more likely motivated in their career choices mainly by nonenvironmental interests.

Of course, most people involved in both governmental and nongovernmental environmental activities are neither scientists nor engineers. In fact, little of the modern environmental agenda has been explicitly set by technical experts. So minorities' low representation in the biological and chemical fields does not in itself explain their low representation in the ranks of the environmental movement. Additionally, this particular finding of the aforementioned EPA survey on stated differences in white versus African American reasons for working at EPA could mean many things, of which lack of minority interest in the environment is just one possibility. Nevertheless, the results of that survey question are troubling.

Whether or not it is the primary cause, many members of the mainstream environmental movement appear to accept lack of interest in the environment as an important cause for low minority representation in the rank and file of government agencies whose mission is the protection of the environment and natural resources. They also assign the same reason to minorities' almost complete absence among both the ranks and the leadership of national environmental nongovernmental organizations. In 1992, the Conservation Fund presented 515 leaders of conservation and environmental nongovernmental organizations with a survey of eleven critical or negative statements. The statement that the professional staff leaders most agreed on was, "Most minority and poor rural Americans see little in the conservation message that speaks to them."[3] In a corresponding survey for volunteer leaders, this statement elicited almost exactly the same response. These leaders—of whom about 38 percent were oriented toward environmental issues versus conservation (36 percent) or preservation (10 percent) —simply did not see their message or orientation as appealing to minorities or the poor.[4]

Another reason often advanced for lack of minority representation is bias within environmental organizations. Chapter 3 in part examined why issues of race and social justice were not included in the agenda of the original environmental movement. One proposed reason was that the environmental movement itself was founded on an antiurban ethos that allowed for little consideration of the plight of the poor and minorities. A natural consequence of such a philosophy was the absence of social justice issues from the original environmental agenda. But the implication of Chapter 3 was that there was no deliberate effort to exclude minorities and the poor from the environmental movement, yet the movement until very recently has remained white and middle-class. To use a phrase heard during cultural diversity workshops on university campuses across the country, the environmental movement, given its predominantly white, middle-class, suburban background, may have created a "chilly climate" of subtle (or not so subtle) racial discrimination that discouraged minority participation. But these are probably not all the reasons for the lack of minority participation in the environmental movement. Remember that in spite of the chilly climate of universities and colleges, since the 1960s, minorities have with some variation increased their representation among university students and faculty. So the question may be asked: why has this not occurred within the environmental movement?

The range of the debate thus stretches from whether minority nonparticipation in the environmental movement is the result of indifference to whether that lack of participation is the result of discrimination (either direct or indirect). A close examination of the debate, however, reveals several information gaps. There is little detailed and focused information— as opposed to anecdotes and opinions—on minority attitudes toward environmental issues. The commonly held belief that minorities simply are not interested in environmental issues may be more a case of shifting blame than one of reality. Conversely, while nongovernmental environmental groups have historically exhibited a high degree of racial or social indifference, there has been no systematic study conducted proving or even strongly indicating deliberate discrimination against minorities. In short, there is little empirical support for either lack of interest or discrimination as explanations for minority nonparticipation in the environmental movement.

Minority Attitudes toward the Environment:
Some Survey Results

The first order of business, then, is to get a clearer idea of the attitudes of minorities toward environmental issues. Unfortunately, there are few large-scale surveys or studies to turn to. Some of the earlier studies in this area agree with the disinterest hypothesis. For example, in the 1970s, M. R. Hersey and P. B. Hill surveyed over two thousand elementary and high school students in Florida and concluded that African Americans are less environmentally oriented than whites.[5] And to L. Milbraith, "it appears more valid to say that many blacks are simply indifferent" to environmentalism.[6]

But for our question, these early surveys had a major design flaw. Most of these early studies tended to focus on minority attitudes toward mainstream environmental groups and their specific agendas. They usually asked respondents whether they supported specific environmental causes within the framework of existing "environmental" organizations. This is very different from asking whether they had an interest in more general environmental issues.

Most of the environmental debate in the modern era has tended to be forced into the parameters of these national organizations and their advocates. But as a group, these organizations represent only a segment of a wider range of environmental issues and ideals.[7] To extrapolate interest or disinterest in environmental issues in general from the goals and activities of these limited issue groups is clearly a mistake. The question of minority interest and commitment thus remains unanswered.

Over a period of several years, the Roper Organization conducted many types of public opinion surveys in which environmental questions often were included as part of a larger set of questions. An examination of the results of some of these surveys could provide answers to the above questions and some differentiation of responses along racial lines. Furthermore, several more recent studies have tried to focus on more general environmental interests rather than on the support of specific items of the environmental agenda.[8] The results of all these surveys indicate that environmental issues are considered crucial by the minority community, and further, that environmental interest is nearly identical for both minority and majority populations. The Roper surveys do suggest, however, that minorities, notably African Americans, may not rank environmental issues as high on their priority list as the majority population does. Other social issues—crime, drugs, and racial discrimination, among others—appear to be

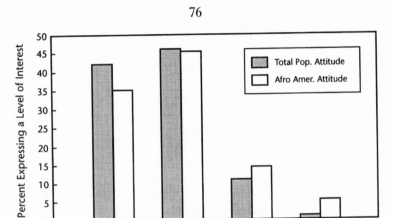

Source: Roper 90-10

Figure 5.1. Degree of interest in environmental issues.

higher priorities for many members of minority populations. But although environmental issue may occupy a slightly lower relative position within parts of the minority community, the surveys indicate that absolute interest in these issues is comparable to that among whites.[9]

At the same time, according to a series of Roper survey questions, minorities feel a greater sense of individual helplessness in the face of environmental threats and are less likely to directly participate in environmental organizations, perhaps more out of frustration than disinterest. Finally, perhaps due to greater economic stress, minority respondents were less likely than the average respondent to be willing to pay for correcting a particular environmental problem.[10]

Consider first the question of whether minority communities view environmental issues as important. In 1990, the Roper Organization conducted a series of surveys asking citizens whether they had an interest in various issues, including environmental issues.[11] Individuals evaluated each issue separately and did not have to rank or compare multiple groups of issues. Unfortunately, Roper divided its survey populations into only "Total" and "Black" (or "African American"), so a broader range of minority attitudes is not present. Of the total sample in the 1990 surveys, 88 percent said they were either "very" or "moderately interested" in environmental issues, whereas 80 percent of the African American population had a similar response (Figure 5.1).

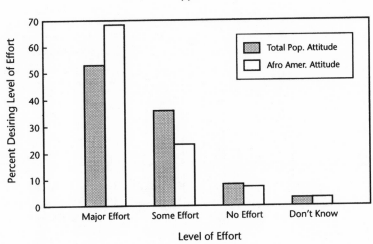

Source: Roper 91-9

Figure 5.2. Level of government effort to solve ghetto, race, and poverty problems.

In a related survey in 1990–1991, Daniel Krause asked residents of the Chicago area, "All things considered, would you classify yourself as an environmentalist?" About 57 percent of all respondents said yes, with no significant variation in response by ethnicity, income, sex, or education.[12] When respondents were asked about specific environmental problems (such as hazardous waste, air and water pollution, the ozone layer, and population growth), the levels of concern were uniformly high, with one exception: population growth.[13]

But an issue may command interest without inspiring the commitment of resources and attention to solve it. In a series of surveys conducted in 1991, the Roper Organization asked respondents how much government effort should be devoted to a variety of problems, such as ghetto, race, and poverty problems, crime and drug problems, and environmental problems.[14] Both the African American population and the total population had very similar attitudes toward nonenvironmental issues (Figure 5.2). In order to solve ghetto, race, and poverty problems, the majority of both the total and African American populations desired a "major effort" from government, although the African American "major effort" response was somewhat higher than that of the aggregate sample.

We would expect African Americans—who, it may be argued, are more immediately affected by social injustice, especially racial problems, and economic hardship—to be more desirous of government help in these areas

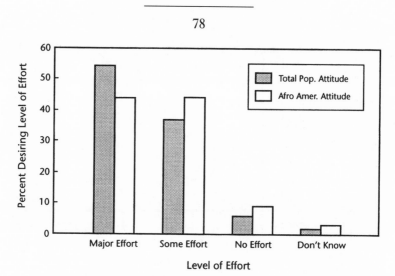

Source: Roper 91-9

Figure 5.3. Level of government effort to improve the quality of our environment.

than the total population. It is interesting that for the issue of improving environmental quality, the African American response was quite close to the response of the total sample (Figure 5.3). These responses clearly discount the notion that minorities do not care about solving environmental problems.

In a similar vein, Paul Mohai, examining a 1980 survey by the U.S. Department of Agriculture and Louis Harris on the public's attitude toward conservation, extracted three factors that determined environmental concern.[15] No statistically significant differences were observed between whites and blacks or across income groups. Furthermore, Mohai controlled for socioeconomic categories such as income, education, and occupation. He was thus able to test the hypothesis that differences in interest between blacks and whites were due to a hierarchy of needs in which environmental issues were of less absolute importance to minorities. That is, even when socioeconomic categorical rankings were controlled for, blacks and whites exhibited the same level of concern toward the environment. Moreover, the hierarchy-of-needs hypothesis did not appear to apply in that environmental concerns remained significant to both races regardless of income.

Similar conclusions about minority interest in the environment were drawn in several other studies. Among them, Judi A. Caron obtained the response of an exclusively black sample to Van Liere and Dunlap's "new environmental paradigm." When she compared these results against Van

Liere and Dunlap's original all-white sample, she found black concerns at an absolute level were similar, although different on some specific issues and on relative priority.[16] Francis O. Adeola's survey of residents in the Baton Rouge, Louisiana, area also found "no significant evidence showing that one racial group is more concerned about environmental conditions than the other."[17]

The picture changes slightly when environmental concerns are compared with other social or economic issues. In a 1993 Roper survey in which respondents were asked to identify the two or three most important issues among a set of twenty, pollution (the sole environmental choice) was a low priority for both the total sample and for the African Americans surveyed.[18] For both groups, the social issues of crime and drugs were high priorities. In contrast, only 15 percent of the total survey group, and 7 percent of the African American population sampled, listed pollution as an issue about which they were "most concerned."

In studies that compare issues, an environmental question could place very low on a list that includes issues such as crime, racial relationships, drug abuse, and unemployment. In this particular Roper survey, the African American response was 50 percent lower than the total survey response on the issue of environmental pollution. But what is important here is that for both the total and minority groups, the pollution issue was important, although not as important, on average, as social issues such as crime and drug abuse. This does not mean that either group does not perceive environmental issues as important. They simply are not as crucial as other issues.

As for a willingness to live near (within ten miles of) certain types of potentially hazardous facilities, the total population and African American responses to a 1992 Roper survey are very similar.[19] As Figure 5.4 shows, for both the total surveyed group and for African Americans, living near a nuclear waste site was very unattractive. Perhaps as striking is that for both groups, landfill sites and large incinerating plants for garbage rated much smaller negative responses.[20]

Before anyone concludes that either the total population or African Americans in particular would want to live near a landfill or an incinerating plant, remember that asking people whether they would prefer such facilities over a nuclear waste site was similar to asking someone whether they preferred being killed by lethal injection or being bludgeoned to death with a club; most people would prefer to avoid either fate. The survey results show only that both populations tend to have relatively similar responses when listing undesirable outcomes. The strong and steady opposition by all communities to the location of landfills and incinerators in their midst attests to the power of these issues, when considered separately, to generate strong feelings. Perhaps the most telling point is not so much the

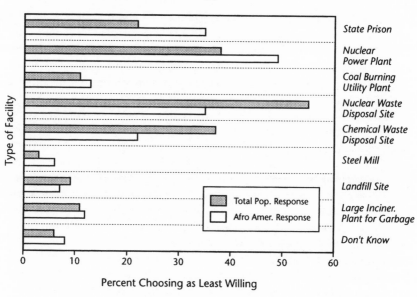

Figure 5.4. Willingness to live near an environmental risk.

relative tolerability of landfills and incinerators as it is the very strong negative reaction by both blacks and whites to nuclear activities.

These surveys strongly indicate that at least the African American population desires government action on environmental issues. In a series of surveys that ask more specifically about the solid-waste problems and the improper disposal of chemical waste in their own community, African Americans actually show greater concern than does the total population.[21] These are not the responses of a group that does not care.[22] And this strong environmental concern among minorities is not a recent phenomenon. In a 1983 Roper survey, when asked their degree of concern over the problem of improper chemical waste disposal, more than two-thirds of the African Americans surveyed were "very concerned"—virtually the same (actually 1 percent higher) as the total population.[23]

If It's Not Minority Attitude, Then What?

Assuming that these survey results accurately reflect the attitudes of minority communities, what explains the virtual nonexistence of minority participation in the making of national environmental policy? What explains the lack of a serious minority presence in any of the major nongovernmental environmental organizations? What explains why the major governmental environmental policy-makers in agencies such as the EPA and the Departments of the Interior, Agriculture, and Energy are predominantly white? Answers to these questions, if they exist, are complex, multidimensional, and involve many more reasons than a hostile environment (both perceived and real) for minorities in virtually all areas of the environmental movement, both in and out of government. Many of these reasons also reinforce each other, resulting in a broad array of forces hindering minority involvement in the movement.

First, there is no question that minorities have virtually no presence in national environmental organizations. Accused of racism—or, at best, racial insensitivity and indifference—the national environmental organizations have only within the last several years begun to act as if a minority voice is not only important in the environmental policy debate, but essential for successful environmental action in the twenty-first century and beyond. In the early 1990s, one could have found few faces of color among professional levels of the major environmental groups. In 1990, of the 315 staff members of the National Audubon Society, for example, only three were black; the Sierra Club had just one Latino on its 250-person staff, and among its professionals, there were no blacks or Asians. In fact, after several environmental-justice and civil rights leaders accused the eight major environmental organizations of racism in hiring, the organizations admitted to poor minority-hiring records.[24]

Mainstream Organizational Attitudes toward Minority Participation

But while admitting a poor effort in the hiring and recruiting of minorities, spokespersons for the groups did not admit racism. Said one, "We are not proud of our record—we are terrible. But I can't believe it is racism. We are not getting the candidates from the minority community."[25] Some leaders of mainstream environmental organizations argue that minority underrepresentation simply results from the failure of their organizations to actively recruit minority candidates and from the failure of environmental issues to attract those minorities who are interested in "causes." They further argue that the relatively low salary packages offered by these institutions are not very attractive to the minority professionals who meet their experience standards.

Several environmental-justice activists offer a slightly different explanation. These activists argue that although mainstream environmental organizations are not guilty of deliberate discrimination, their pursuit of business as usual virtually guarantees limited minority recruitment. First, the major environmental organizations recruit primarily from predominantly white universities and colleges and from programs in environmental science, natural resource management, and the biological and physical sciences—fields with below-average minority representation. These environmental groups still have no major plans to contact science or environmental studies programs at historically African American or Latino institutions. Nor has there been, until recently, any attempt to examine the logic of their rather narrow avenues of recruitment. Second, whether deliberate or not, these organizations have until very recently made little or no attempt to portray their mission in a manner more attractive to potential minority candidates. From their literature to their published agenda, the image they projected was that of a white face pursuing beautiful green fields and warm, fuzzy objectives.

This failure to attract minorities also reflects the differences in mission focus between mainstream national environment groups and many grassroots environmental organizations. The former tend to emphasize lobbying, legislative solutions, and the increased professionalization of staff ranks. The grassroots organizations favor direct action, citizen empowerment, and the extensive use of volunteers.[26] The national organizations' agenda may also be influenced by the reliance of many of them on foundation and corporate funding versus the membership dues–based operational funding of most grassroots groups.[27]

Attitudes and priorities are important in increasing involvement of minorities in national environmental organizations, and the attitude of mainstream environmental–preservation–conservation organizations is not good. In fact, Robert Bullard and B. Wright argue that the antiurban bias prevalent in the leadership of these organizations is a major stumbling block to greater minority participation.[28] Consider the results of a survey by the Conservation Fund in which 515 leaders of environmental and conservation organizations were asked to prioritize seventeen issues and programs.[29]

Of course, this survey does not prove that nonwilderness issues have a lower status among minorities. The survey group was composed of preservation–conservation and environmental leaders. So the preference for nonurban themes of natural resource management and preservation–conservation should not be surprising. There was unfortunately no separate analysis of the "environmental" groups alone. Nevertheless, it is surprising that even among this mixture of leaders, environmental themes rank so low.

What is of more concern is that many, if not most, of the senior environmental policy-makers in government come from these same environmental or preservation–conservation organizations. These organizations also provide the major recruiting pool for career environmental professionals in government. So given the absence of minorities from the staff and professional levels of these nongovernmental organizations, in essence, government recruitment of environmental policy-makers and career professionals is effectively biased against minorities. A major voice, and a different voice, in the environmental argument has thus been held silent in government circles. Lacking this "other" dimension, the response of government agencies, which affects all citizens both in and out of environmental movements, may fail to properly address environmental problems.

Although the mainstream environmental movement is quite diverse in its goals and philosophies, all groups share parts of a basic belief system. Unfortunately, this common belief structure (as examined in Chapter 3) until very recently had no place for issues of unequal environmental impact. Amazing as it seems from this side of the discussion, until very recently, the question of the fairness of environmental and preservation–conservation efforts to specific groups of people simply was not seriously discussed within mainstream environmental organizations. And when it finally was discussed, the subject arose not from the sudden enlightenment of some cohort of these groups in or out of government, but rather from the shouts of grassroots groups of minorities and other local activists as they demanded a hearing.

In this case, the common policy on recruitment sources, which draws from a rather limited range of possibilities, regardless of intentions, has caused agencies to be blind to major aspects of the environmental question. Thus, although it appears that minorities are interested in environmental issues to about the same degree as others in the population, it does not follow that they would be interested in the same environmental problems as—and participate in large numbers in—mainstream environmental or conservation groups. But without that connection to mainstream environmental or natural resource organizations, access to positions in governmental agencies has been severely restricted. And without some representation within these agencies, it is difficult to effect the modifications in agenda necessary to make environmental activism more attractive to minority populations. We thus have a vicious circle.

But most of the blame for this condition cannot be placed on mainstream environmental groups. If minorities perceived environmental issues as sufficiently important, then organizations directed toward the environmental concerns of minorities or social ethics environmental concerns should have appeared earlier, as they did for civil rights, crime, and poverty. It does not take membership in a mainstream environmental group to become involved in environmental activities. Why in the 1960s and 1970s did organizations not emerge to rival the mainstream environmental groups and create their own agenda? For example, from 1964 to 1992, the annual reports of the NAACP Legal Defense Fund made no mention of inequities in environmental impact—such as the need, in some areas of the country, to challenge the disproportionate location of hazardous facilities in predominantly minority communities.[30] The nearly complete absence of environmental issues from the agenda of this social action organization parallels the absence of social justice or ethics issues from the agendas of mainstream environmental organizations.

More than just the chilly climate of many mainstream environmental organizations may have dampened minority interest in the environmental movement. The lack of concerted minority involvement in these environmental-justice issues suggests that dynamics within the minority community may have also played a role.

As explained in Chapter 4, the evolution of environmental justice within the minority community required a shift in minority understanding of environmental issues. The almost complete lack of commentary about environmental issues by most minority and social justice leaders strongly suggests an inability to connect such issues directly to minority life and welfare. And once again, given the wilderness orientation of most environmental organizations until the late 1980s, it should come as no surprise that minority organizations saw little in the stated goals of the mainstream environmental movement that spoke to their needs.

Regardless of race, many individuals may express interest in an issue, but far fewer will actually participate in activities related to that issue. Environmental issues are no exception. In 1989, a Roper survey asked individuals if they regularly worked for a local environmental group; only 3 percent of the total group and 1 percent of the African Americans said they did.[31] Both response levels are small, but the total population response was three times greater than that of the African American respondents. What is more telling is that although 20 percent of the total group admitted that they "contributed money to an environmental group," either regularly or from time to time, only 7 percent of the African Americans responded that they did so.[32] Obviously, the minority influence on the agenda of environmental groups, both in terms of participation and financing, is very small.

The cause of this outcome could be circular. Minorities do not participate or give money because they find little of interest to them in the current agendas of most environmental groups. But most environmental groups do not pursue goals more attractive to minorities because minorities have such little input in the decision-making process of these groups.

There could, of course, be many other reasons for this difference in attitude toward the power of the individual to bring about change. If minorities could feel more able to change their environment, then their noninvolvement in environmental policy activities could change. Part of that change is the emergence over the last decade and a half of many community-based environmental groups that have not only recruited a large minority membership, but have also established the beginnings of a national network. The activities of these groups could have a profound positive effect on the minority community's sense of empowerment in environmental policy and outcomes.

What's Next?

Two of the misconceptions alluded to in the title of this chapter are that the minority population has no interest in environmental issues and that the only way environmental interest can be shown is through participation in the mainstream environmental movement or by following an agenda modeled after the mainstream. Many mainstream environmentalists may still need to believe in an uninterested minority population as an excuse for their organizations' inequitable membership, both now and in the past. But the fact is that minorities have as much interest in environmental protection as the majority population—and on specific issues, perhaps more. They may place certain other social issues above environmental concerns, and they may prefer a different ranking among environmental issues than the current membership of many mainstream environmental organizations, but differences in priorities do not mean lack of interest.

Although minority membership in mainstream environmental organizations may increase, it does not appear likely, given minorities' fundamental differences with the majority over agenda, that minorities will be joining such groups in proportion to their numbers in the population. As several of the surveys cited above make clear, although virtually everyone "supports" environmental protection and corrective action, only a small percentage of either majority or minority populations will ever actually be active in the movement. But it is by those actively involved that much of this country's environmental policy is determined. For that reason, we need to be concerned about the proportional participation of all segments of our society.

It is in the interest of mainstream environmental organizations to increase their minority membership and leadership. However, these organizations have the right to form their agendas as they see fit and to define *environmental interest* as they desire. But, given the current bias of their agenda, what cannot be permitted is that they continue to serve as virtually the sole source of recruitment for government agencies. Agencies with mandates that apply to all citizens have no right to eliminate opportunities for different voices to be heard within their own numbers. This is not a case of environmental affirmative action, but rather of eliminating an unfair hiring approach that has far less to do with realistic needs than it does with habit.

Six

The EPA
An Agency with an Attitude

Environmental justice is not a concept that has come easily to the environmental and natural resource (ENR) establishment. The concept flies in the face of the very ethos upon which much of the environmental movement was founded and to which many of its members still subscribe. Only after considerable prodding has the U.S. Environmental Protection Agency (EPA) conceded that problems of differentiated environmental impacts deserve consideration in the policy-making process. But it is likely that many of the EPA's rank-and-file professionals—or those of any other ENR agency, for that matter—give little more than lip service to the concept of environmental justice.

In the last three chapters, we have seen how environmental justice emerged as a viable social issue. We have also seen why social justice and race were not part of the original agenda of the early environmental movement. And we have examined the complex issue of minority attitudes and participation in the environmental movement. In this chapter, I will explore these themes from the governmental side of the equation. As noted in Chapter 4, one of the major problems with the current governmental agenda is that many of the people who formulate and administer it come from similar backgrounds in the nongovernmental environmental movement—an area in which sensitivity to social justice is not well developed. This chapter focuses on the restrictive attitudes of the federal agencies whose missions include the environment and natural resources—and in particular, on the attitudes of the EPA.

Of course, by no stretch of the imagination can or should it be argued that the attitudes of the personnel of one federal agency, or even a collection of environmental and natural resource (ENR) agencies, dictate the direction or policies of the environmental movement in this country. Even literature on the most proactive policy formulation concedes that governmental agencies play at best a major role—but never a solo role—in the formulation of policy. On the other hand, the belief structure and behavior of members of a particular collection of governmental agencies provide useful insights into the difficulties and probability of acceptance that the policy initiatives related to environmental justice will encounter as they evolve.

In this chapter, I examine at length the results of a personnel survey and employment census conducted at the EPA in the early 1990s, just as the environmental-justice movement was beginning to have an impact within the federal system. The survey results suggest that a sizable portion of the EPA's white-majority ranks may have had a serious problem with issues, such as environmental justice, that involve social and cultural sensitivity. The survey investigation is followed by a review of personnel patterns from the EPA and other federal ENR agencies over the ten years from 1982 to 1992, when environmental-justice issues were first gaining national prominence. Employment patterns, analyzed by government service (GS) rankings and race, reveal that this cluster of agencies is somewhat different from the federal norm but similar to the nongovernmental ENR organizations from which most of their membership is drawn.

What is equally interesting is that according to the survey, during the time period covered, various populations within the EPA, especially white men when compared with minority men, differed radically on a number of key issues of organizational life. Those issues included why they were working at the EPA, how fair they felt the EPA was to employees, and how good the EPA's minority hiring record was.

What Attitude?

This difference in attitude also extends to policy issues, particularly issues of environmental-justice policy. For example, an employee in the EPA's senior executive service (SES[1]) with both regional and headquarters experience, when asked in 1993 about the problem of environmental justice, began by describing the issue as "a bunch of crap." He went on to argue that investigation of the issue was a waste of valuable resources, merely a response to uninformed activists with a misguided sense of injustice or injury. Although no formal poll on environmental justice has ever been taken among EPA personnel, there is the suspicion that an anonymous survey of more senior EPA professional-rank employees (those with the rank of GS-14, GS-15, and SES) would discover that the sentiment stated above is shared by more than a few senior professionals within the EPA. There is also a suspicion that a similar survey conducted among professionals in the major nongovernmental ENR organizations—such as the Sierra Club, the Wilderness Society, and the Audubon Society—would have produced the same results. In the same manner, at federal ENR agencies other than the EPA, such as the Departments of Interior and Agriculture, the question would elicit the same disapproval—or worse, a blank stare.[2]

This portrait does not ignore the official EPA effort, especially within the last decade, to embrace, however tentatively, the issue of environmental justice. At the beginning of the Clinton administration, the EPA had identified as one of its top priorities the addressing of environmental inequities. By the late 1990s, hundreds of employees at both the headquarters and the ten regional offices of the EPA were devoting a significant portion of their time to analyzing, structuring, reviewing, and proposing plans on questions of environmental justice. The EPA has progressed from having an Environmental Equity Workgroup in 1990, to a small Office of Environmental Equity in 1992, to a still small but potentially growing Office of Environmental Justice in 1995, to finally establishing in 1998 an Office of Civil Rights to handle complaints regarding Title VI of the Civil Rights Act of 1964. Furthermore, in 1993, by federal charter, a National Environmental Justice Advisory Council was established to "provide independent advice, consultation, and recommendations to the Administrator of the U.S. Environmental Protection Agency . . . on matters related to environmental justice."[3] The official agency line is that environmental justice is an important issue, one that all EPA employees must support and address in carrying out their program missions. The EPA now portrays environmental justice as a subject that must be integrated into all aspects of EPA policy analysis, policy-making, and program implementation and operations. One still waits for a clear sign that the EPA has made a serious policy-level commitment to environmental justice—a commitment beyond lip service and a few token appointments—or whether the EPA has the appreciation of the dynamics of social justice and differentiated impact to make reasonable policy recommendations. And even if there is such a commitment on the executive level, how dedicated to such a philosophy and policy agenda are the rank-and-file professionals who will be responsible for carrying out those policies?[4]

Equally uncertain are the attitudes of those now entering the environmental field, especially those who will make up the next generation of EPA senior professionals. Consider the all-too-common reaction of graduate students taking an environmental policy analysis course in the spring of 1998 in a school devoted to improving environmental management and policy analysis. When forced to confront the topic of environmental racism, the students, many of whom had had several years of organized experience in environmental protection, spent more time defending the agenda of the traditional environmental movement than assessing the issue objectively. As with the senior EPA professionals, the students constantly attempted to devalue this discussion; finally, they dismissed it as inappropriate for environmental studies. Ironically, students who always complained about the unfairness of market economics when dealing with other environmental problems were quick to arrive at a consensus that the environmental-justice problem was less one of environmental discrimination and much more likely just an example of those same market forces—now suddenly "fair"—at work.

Why are environmental professionals unwilling to seriously consider that the current patterns of environmental risk distribution have more to do with race and ethnicity than chance or even economics could predict? Why was the General Accounting Office, ensconced in the legislative branch of the federal system and not normally thought of as a bastion of progressive thinking, years ahead of the EPA in arguing that race was a significant factor in the distribution of certain toxic-waste sites or other hazardous conditions?[5] Why have the other major ENR agencies such as the U.S. Departments of Interior, Agriculture, and Energy lagged even further behind the EPA in even recognizing that a problem exists?[6] Why during the last ten years has the EPA devoted virtually no research dollars to systematically. examining the various dimensions of this question? In fact, why has the EPA Office of Research and Development, until a few years ago, had a nearly blank record of supporting research on the social impact of environmental issues?[7] Such neglect surely cannot have occurred because the issue of environmental justice became serious only within the last few years. The problem of differentiated population impacts on policy decisions and activities has existed since the inception of the environmental movement.

How does one explain such extraordinary and prolonged insensitivity and inaction on social policy on the part of the EPA and the other ENR agencies before the 1990s?[8] To answer that, let us return to the question asked in several earlier chapters: Could all these governmental institutions be fundamentally racist in their policies and beliefs? If not, what institutional traits could explain a national condition in which many minority communities are exposed to disproportionately high levels of environmental risk, as well as a slower and weaker governmental response to such risk? And while such conditions exist, an EPA professional can still declare that issues of environmental justice are "a bunch of crap."

As previously noted, the EPA still is dominated, regardless of the party in power in the White House, by professionals who come from a culture that is wanting in sensitivity to social justice. To incorporate a concept such as environmental justice into the EPA's operational reality, its leaders may have to first recognize the internal institutional and cultural barriers working against this change.

In this discussion, the EPA is not the exclusive target. The conclusions drawn here about the EPA apply, to an even greater extent, to the other ENR agencies. Problems of sensitivity to social justice apply at least as strongly to agencies such as the U.S. Departments of Energy, Agriculture, and Interior. The EPA just happens to be far enough along the path of incorporating environmental justice into its policy structure to warrant some observations. The EPA gets the spotlight here—not because it is the worst example, but rather because the EPA is most closely identified with environmental policy and, perhaps more important, because attitudinal information on EPA personnel is available.

Organizations: Everyone Does Not Agree

All organizations, public and private, have cultural norms and traditions. In the case of the environmental movement, these norms and roots virtually guarantee that issues of social policy, particularly racial policies, receive little attention and hold relatively little interest for most of the movement's participants. The environmental movement—especially the national environmental movement—has been and remains an essentially white, middle-class culture based on suburban values. Growing out of the natural resource and conservation movement of the late nineteenth century, the environmental movement had room for the needs of the individual, particularly the nature-appreciating individual, but oddly had little or no room for the needs of specific groups of individuals, particularly groups of poor or minority individuals. Actually, in many ways, the environmental movement focuses on things, not people. The extreme example of this attitude are the animal rights activists who hold the life of an animal—any animal—equal to that of a human. This attitude is also evident in the proliferation of collegiate summer internship programs offered by so many universities and nongovernment organizations that allow students to spend several months in a developing nation, studying the impact of industrialization or urbanization on indigenous flora and fauna; while on the other hand, there are few summer intern programs inviting these same students to help resolve the environmental problems of poor communities in their own backyard.

Although many federal and state agencies have some responsibility for environmental protection and management, the lion's share of the protection mandate resides with the EPA. It is within the EPA that the hazardous-waste site identification process (Comprehensive Environmental Response, Compensation, and Liability Information System) resides; it is the EPA that has responsibility for overseeing the Toxic Release Inventory reporting system; and it is the EPA that has executive primacy for most major environmental protection legislation, such the Clean Air Act and the Clean Water Act. Thus, as a way of appreciating the culture in which issues of environmental justice have held such a low priority (to the point of disbelief), we begin our scrutiny of environmental agencies with the EPA.

Some advocates of environmental justice, as well as community activists, charge that the EPA has never been a friend of minority communities, has been insensitive to problems of the poor, and has taken a much less aggressive approach in addressing pollution problems found in minority and poor communities than those found in white communities. It has further been charged that the agency's professional workforce has almost no cultural sensitivity and little, if any, appreciation of the community dynamics of environmental risk in poor and minority communities. In this chapter, I will try to determine whether these charges are true.

The EPA was created December 2, 1970. Its mandate and personnel were pieced together from programs and divisions of various other agencies. From the old Department of Health, Education, and Welfare (HEW) came the National Air Pollution Control Administration, the Bureaus of Water Hygiene and Solid Waste Management, and parts of the Bureau of Radiological Health; from HEW's Food and Drug Administration came partial responsibility for management of pesticide levels. The Department of the Interior passed on to the EPA its large Water Quality Administration, and from the Department of Agriculture came an additional part of the EPA's pesticide mission, the final step in pesticide control: pesticide registration. Finally, from the Atomic Energy Commission came the authority over radiation standards.[9] The majority of the transferred personnel were either scientist-engineers or regulators. Neither group had had much experience dealing with social policy, nor did they have any particular mandate during EPA's infancy to shift their focus in such a direction.

According to the published oral histories of the first two EPA administrators, William D. Ruckelshaus and Russell E. Train, and the writings of some chroniclers of that period, the EPA was distinguished as an agency with committed personnel who shared an ethos and a purpose.[10] Made up primarily of college-educated white men with backgrounds in science, engineering, or law, there is no question that most of these early EPA pioneers were committed to addressing environmental problems. At the time, Ruckelshaus stated, "There are no finer public servants anywhere in the world than the men and women of the EPA."[11] It is also clear that their naiveté about economic effects systems and industrial management gave many of them a false sense of how quickly they could "solve" the problem of environmental pollution. But nowhere in that early ethos is found any concern over the disparity of environmental impact across social or racial groups.

Russell E. Train, the second EPA administrator, made this view quite clear when he identified EPA's primary constituencies as businesses, environmental organizations, and farm groups.[12] No mention was made of minority or low-income communities, which might suffer the greatest environmental burdens, or of classes of citizens, such as farm workers or consumers of large amounts of fish, who are exposed to levels of particular environmental hazards that are significantly higher than normal. The agency's emphasis was on improved regulatory enforcement and the expansion of both basic and applied research on environmental pollution.

But the EPA was not the only agency largely filled with dedicated people who had a relatively weak sense of the social dynamics of their mission. This same description can be applied to the other large ENR agencies, such as the U.S. Departments of Interior, Agriculture, and Energy. The poor record of the EPA and these other agencies in the area of environmental justice is not so much the result of deliberate obstruction of justice or planned discrimination as it is the simple result of a culture that does not place much emphasis on any social policy, particularly as those policies relate to urban conditions.

The urban environment is not where nature rules. Urban problems do not have quite the cachet of environmental problems that affect "natural environments." Until recently, for instance, the emphasis in the controversy over acid rain was not on how it affects urban structures or residents, but rather its effects on forests and lakes. Problems of deforestation appeared more immediate and threatening than did the problems of the urban poor. From this perspective, it should not be surprising that many perceive the EPA professional as not very sensitive or responsive to the environmental plight of minority communities. In fact, the surprise is that in spite of a history of urban and social insensitivity, the resistance to the incorporation of even the modest current environmental-justice initiatives and strategies at the EPA has not been greater.

Survey of Cultural Diversity

A number of assertions about EPA employee attitudes and orientations have been posited here as evidence for broader social insensitivity among ENR agencies in general. But perhaps all this is the misguided musings of critics, musings based on little more than the quotations of a few administrators and an occasional anecdote from a graduate class or two in environmental policy.

In the spring of 1992, the EPA conducted a cultural diversity survey of all employees in its Washington, D.C., headquarters. The final report on the survey, released in June 1993, presents a number of findings about employee attitudes toward working conditions in the EPA and related attitudes toward issues of social policy during a critical period in the development of an EPA environmental-justice strategy.[13]

It is a principle of consumer choice theory that what individuals *say* they prefer or do may be very different from what they *actually* prefer or do. Note, for instance, the high incidence of moral indiscretions among self-proclaimed television evangelists. Responses are generally limited, even in blind surveys, to what is perceived as socially acceptable.[14] Especially with attitudinal surveys, respondents are not so much guilty of conscious distortion as they are of unconscious self-censoring or bias.[15] Yet in spite of this handicap, survey information at least affords a significantly more structured and reliable examination of organizational attitudes than do simple anecdotes and comments. Short of examining in detail the actions of every employee, a survey of EPA employee attitudes provides useful insights into why the concept of environmental justice first met (and in some quarters of the EPA, still meets) with disbelief and resistance.

What is interesting about the results of the EPA survey is that despite the very likely difference between revealed and stated preferences noted earlier, respondents to the survey gave enough information to suggest that problems of cultural and racial sensitivity should be of real concern at the agency. Assuming that all respondents had some idea about what the "acceptable" responses should be, the simple fact of ethnic group differences in response within the EPA argues that the severity of the problem may be even greater than evidenced by the survey. While not explicitly focusing on the issue of environmental justice, the questions in the survey reveal a pervasive cultural insensitivity or indifference in many EPA employees.

The sixty questions in the survey covered a broad range of cultural diversity issues, such as perceptions of the EPA's recruiting, training, supervising, and promotion record, and the degree of cultural sensitivity at the agency, as well as general issues such as reasons for working at the EPA and general job satisfaction. The survey, part of a three-stage effort to study cultural diversity, highlighted several key areas where clear differences existed across racial and cultural groups concerning conditions at the EPA.

Consider, for example, perceptions about the EPA's record of minority recruiting.[16] When presented with the statement "the EPA has a good record of actively recruiting minorities," the majority (72 percent) of white respondents agreed, whereas most African Americans and Hispanics disagreed. Likewise, 70 percent of whites agreed with the statement "Affirmative action and equal employment opportunity programs at the EPA have been successful," whereas a majority of all minority groups disagreed.[17] When asked whether "Affirmative action policies at the EPA lead to hiring less-qualified employees," a majority of all groups disagreed; but when broken down by sex, the results indicated that a majority of white men agreed with the statement. For this last statement, the level of agreement that EPA affirmative action policies led to hiring less-qualified employees increased as the education level of the respondents increased.

In light of these results, whites at the EPA, especially white men, appeared to believe that affirmative action and minority recruiting had been reasonably successful, whereas many minorities had just the opposite attitude. Worse, white men in particular held the belief that such programs resulted in the recruitment of less-qualified individuals. This belief becomes more significant when one realizes that in 1992, whites made up nearly 74 percent of the total EPA workforce of 18,599—and, even more significant, comprised 95 percent of the executive ranks and 84 percent of the more senior GS ranks (levels GS-12 through GS-15).[18] In short, most whites at the EPA seem to have perceived the EPA's history of hiring minorities as satisfactory, and most minorities, especially African American men, seem to perceive that same history as unsatisfactory.[19]

With the survey statement "Minorities have equal opportunities for advancement within the EPA," more than two-thirds of whites agreed, while sizable majorities of all minorities disagreed (the tiny sample of Native Americans was evenly split). Even more telling, 77 percent of white men agreed with this statement while 81 percent of African American men, the largest other male group, disagreed.

Similar responses were recorded for the statement "Minorities are treated fairly when it comes to promotions into supervisory and management positions at the EPA," with white men agreeing more readily than any other group. Finally, the survey asked whether "lack of cultural diversity is a source of tension between professional and support staffs." Considering that most of the minorities in the EPA are support staff members, it is hardly surprising that a majority of whites disagreed (with white men disagreeing in overwhelming numbers), whereas large majorities of minorities agreed.

Moreover, as the education level of respondents increased, so did their perception that things were going well at the EPA. But the gulf in approval of the agency was greatest between white and African American men and least among the various minority groups. No questions directly addressed employees' attitudes toward environmental-justice issues, but their attitudes toward conditions within the EPA strongly suggest that most of the agency's white employees would have difficulty accepting the seriousness of environmental justice as a policy issue.

Among the personnel at EPA headquarters, the survey found serious problems of perceived racial and sex discrimination, and a considerable lack of sensitivity, especially among white men, toward issues of cultural diversity. Absent from this or any other recent EPA personnel survey were any direct questions about environmental justice or about social justice in general. But the survey's findings do suggest that the EPA was an organization that by its very nature would not be sensitive to the social policy problems related to its mission of environmental protection. Thus, it should not be surprising that the EPA has been so late in recognizing the existence of such problems and responding to their many ramifications. Furthermore, the actual administration of many EPA initiatives is carried out by its regional and field offices. Anecdotal evidence suggests that personnel at the regional EPA offices are even less sensitive to social issues than are the personnel at its headquarters.

However, assigning blame to the EPA is not the goal here. Consideration of the survey results is intended to indicate only that the EPA has an organizational culture in which sensitivity to social policy has a relativity low priority, and that the white majority within the EPA approach their jobs and workplace with a different attitude from that of their minority coworkers, and quite possibility have little capacity to empathize with the plight of minority communities or citizens.

A further example of this insensitivity is that for the survey question "Why did you come to work for the EPA? To help the environment?", 71 percent of white respondents said yes, but only 42 percent of African Americans did. This is less an issue of differences in commitment to environmental protection than perhaps differences in priority. That is, issues such as job security may occupy a higher priority with minority personnel. When asked, "Why did you come to work for the EPA?", the majority of African Americans appeared motivated by broader issues, such as "security of a government job" (64 percent of African Americans agreed, but only 36 percent of whites); "good benefits" (67 percent of African Americans, 39 percent of whites); or "good working conditions" (50 percent of African Americans, 18 percent of whites).[20] These results correlate with the suspicion, not yet formally studied, that many of the white professionals in the EPA come from the mainstream environmental movement while many of the minority professionals do not.

However, we must not place too much emphasis on the results of the survey summarized above. Given the absence of any equivalent attitudinal survey of any other federal agency, there is no way to easily determine whether the attitudes of many EPA personnel were or are significantly different from those of any other federal agency.[21] Nevertheless, even assuming that similar attitudes are held by other agencies does not diminish the fact that significant attitudinal differences between minorities and whites do exist at the EPA.

Who Works Here?

The differences between white and minority attitudes become more critical when one considers that the EPA has a lower percentage of minority professionals than the average for the federal executive branch. Even more interesting, the federal ENR agencies in general have an exceptionally poor record of minority employment in the professional ranks. In fact, among the federal ENR agencies, the EPA actually has one of the best records. Consider some of the few available records on employment in executive branch agencies, broken down by race. Covering three separate years—1982, 1988, and 1992—the records clearly show the differences between the natural resource and environmental agencies and the federal norm.[22]

In 1982, near the beginning of the Reagan administration, the EPA had a total minority employment percentage very close to the federal average. However, when the lower GS grades are eliminated and only the more senior professional ranks (GS-11 through GS-15 and the SES grades) are compared, the picture changes noticeably. In fact, as the GS grade increases, minority employment at the EPA becomes significantly worse than the federal average. At the GS-15 and SES level, the EPA had a minority percentage 40 percent lower than the federal average. And at the other ENR agencies, conditions were generally worse. In fact, the minority-percentage averages for this group of agencies would have been worse yet were it not for the large number of Native Americans associated with the Department of the Interior. Further, the EPA had a higher percentage of African Americans in every professional grade than the average for the ENR group. To their credit, the EPA, even in 1982, was the only ENR agency above the federal average. But that achievement seems considerably less impressive when one realizes that most minorities at the EPA work in its relatively large support staff, not its professional staff.

By 1988, the situation had not changed very much. The group of ENR agencies was still below the federal agency average both for total minority percentage and for the professional ranks. And as in 1982, as the pay grade increased, the minority percentage decreased. One minority EPA senior professional called this a "plantation" employment structure—a large minority percentage in the lower grades and increasingly fewer minorities in the upper grades.

On the positive side, the ranks of the middle-level professional (GS-11 through GS-13) at the EPA were actually better than the overall federal average. Increased efforts begun in the early 1980s to improve at least the African American presence in the EPA professional ranks seemed to be paying off by 1988. For Hispanics, there was also some improvement at the EPA, but given the initially small Hispanic representation there, it is difficult to determine whether that improvement was intentional or random. For Hispanics, the EPA actually lags behind not only the federal norm, but even that of the other ENR agencies.

By 1992, the situation had continued to improve, at least for the EPA. By then, the EPA's aggregate percentages for African Americans in the GS-11 through GS-15 ranks were higher than the federal average. For Hispanics and Native Americans the numbers were still disappointing at the EPA, whereas Asians, as in 1982 and 1988, were well represented in the middle-level professional ranks. Numbers of African Americans and Asians in the very senior GS ranks were still below the federal average. For the other ENR agencies, the overall situation had not appreciably changed since 1982 (except for the Department of the Interior's large number of Native Americans).

During the four years of the Bush administration (1989 to 1992), minority participation at the EPA and several other ENR agencies improved markedly. In 1992, the EPA's total minority percentage nearly reached the federal average. More significant, much of that increase occurred in the professional ranks. The percentage for GS-12 through GS-15 was virtually the same as the federal average; and the percentage of minorities in the EPA's SES ranks surpassed its 1982 level (which had dipped slightly in 1988). It is interesting that in 1992, minority employment in the GS-12 through GS-15 ranks was the highest among the ENR agencies, but its SES percentage was the lowest. This suggests, among other things, that within the EPA, minorities during the Bush years made percentage gains in some professional grades but were not moving into the senior policy and management ranks.

The numbers strongly suggest that the EPA has definitely made some progress, at least in terms of drawing more minorities, especially African Americans, into the professional ranks of ENR agencies (although this progress has not extended up to the highest professional ranks). Relative to the federal average, the percentages of minority employment in those agencies improved steadily from 1982 to 1992. But while some improvement has occurred for the aggregate of all four of the ENR agencies, the overall pattern has changed very little.

The Future for the EPA

For the EPA, the future is both bright and filled with the winds of change. In the short run, the current culture of the EPA is to a large extent a product of the early form of the environmental movement. Like the movement itself, the EPA must, and appears to have begun to, change to accommodate a different environmental reality. As a part of that change, many of those now at the EPA may have a very different future there. This is a simple recognition that individuals dedicated to an environmental policy predicated on treating physical conditions may find little comfort in environmental programs in which the focus must be on people. And by people, I do not meant some single amorphous mass, as too many environmentalists treat them, but rather individuals and groups that exist under diverse circumstances.

The EPA's current organizational structure (along media-specific lines) will almost certainly become a thing of the past.[23] There were good historical reasons for the current form of the EPA, just as there are now powerful and growing forces and reasons that will push it into its next metamorphosis. But by its very presence, the current structure now hinders rather than aids the EPA in its mission.

There are many in the EPA today who have no idea how radically their careers must change, even in the near future. Many of these EPA employees have adopted at best an air of indifference, at worst one of disdain, toward the question of environmental justice. They fail to realize that this question is not merely a single issue, but rather the harbinger of an entirely different way of perceiving and conducting the business of environmental policy.

As with most government social programs, the EPA will not lead this change but rather follow in the rough waters of its wake. The transition, however, must occur if the EPA is to retain its relevance as the chief environmental voice of the government. But lacking a history in explicit social policy consideration and formulation, the EPA will not find it easy to redefine some of its basic operating principles. Many within the agency will fight against this change, arguing that its mandate has always been the investigation and regulation of environmental hazards, and that the implementation of social policy should be left to agencies such as Health and Human Services, Housing and Urban Development, or the Social Security Administration. They are correct and wrong at the same time. The human element has not been an explicit part of the EPA's mandate, but any implementation of policy carries with it a social policy dimension that will not go away, even if ignored. The division between technology policy and social policy is an artificial one, and the EPA no longer has the luxury of pretending it exists.

As the EPA currently attempts to confront the question of environmental justice, that very confrontation reveals a great deal about missed signals. The agency has an Office of Environmental Justice (OEJ), whose primary mission is to serve as an information conduit, providing both agency personnel and community groups with information on the state of environmental justice. The OEJ was originally placed under the authority of the Assistant Administrator for Human Resources and Administration, a curious organizational location for an issue-oriented office.[24] Today, the OEJ has been moved to the Assistant Administrator for Enforcement, a move that reflects some appreciation of the role the OEJ should play in the EPA. Still, the OEJ has absolutely no program authority, nor does it have the power to redirect or regulate. The OEJ has a permanent staff of fewer than twelve, with little analytic support and virtually no funding resources to allow it to obtain such support from outside the agency. The task of analyzing the problem of environmental justice is left to the guidance of other programs within the agency.

Currently, the EPA is still at the stage of educating its own personnel about exactly what environmental justice means, but the process has been agonizingly slow, even by the glacial standards of bureaucracy. More than twelve years have passed since the original General Accounting Office report citing unequal siting burdens, and more than seven years have passed since the EPA's own major study yielded similar conclusions. Yet while there are directives from senior administrators informing every federal program and group that environmental justice is important, there is little in the way of concrete effort to address the fundamental culture within the EPA, which mitigates against accepting both the seriousness of the problem and the necessity for addressing it.

By following rather than leading, the EPA focuses its information-gathering and -dissemination efforts on the more obvious problems of site location, permitting, and cleanup, giving a minimum of consideration to the possibly graver issues of pesticide exposure, lead poisoning, and even inadequate diet. On the positive side, as the personnel statistics above hint, the diversity in the professional ranks of the EPA, while still not up to the average for all federal agencies, has clearly improved over the last decade. The EPA is perhaps the first of the ENR agencies to reflect, in its increased minority membership, the higher priority that environmental issues now command within minority communities. This increasing diversity may also indicate that compared with the other ENR agencies, the EPA has made a greater effort to recruit and nurture minority professionals—and that the imperative changes in policy may not be as painful as some predict.

If the EPA Has Troubles, What about the Other ENR Agencies?

The EPA may not be perfectly prepared to confront the environmental-justice question, but compared with the other federal ENR agencies and most state operations, it is a shining example of progress and hope. Many of the other ENR agencies appear to have utterly no understanding of environmental justice or social impact, much less any plan beyond the requisite showpiece programs. Worse, beyond the ENR agencies, other federal and state agencies whose activities have environmental impact (and most agencies do fit into this category) are still clueless, still unable to truly appreciate that they have environmental responsibilities of any kind, much less environmental-justice responsibilities.

Very likely all this will change, but not very rapidly or evenly. The Presidential Executive Order of February 1994 created an interagency task force to first bring all federal agencies into the circle of environmental-justice awareness. However, to date, there is little evidence, beyond an obligatory acknowledgment and the obligatory individual agency report, that many of these agencies really see environmental justice as a moderately important (much less a high priority) issue for their own organizations. No one should expect the State Department to appreciate and embrace the issues of regional environmental injustice when they are more worried about regional political and economic collapse. The Department of Commerce certainly does not view environmental justice as a component in their international trade negotiations. Nor does the Department of the Interior's National Park Service concern itself much with the relative underfunding and second-class status of their historic urban sites—which are more immediately significant to urban populations—when they risk losing their large land-preserving parks, whose client population is predominantly middle class.

Part II
Policy Analysis of
Environmental Justice

In any lengthy discussion of a major socioeconomic policy issue, such as environmental justice, at some point one has to deal with what government's role should be. As important as it was to appreciate the dynamics of the evolution and process of environmental justice, it is equally important to carefully consider what role, if any, government can and should take in confronting this issue and the specific form of that role. Not all socioeconomic problems require a governmental solution. An important policy-analytic achievement is determining when *not* to call in the government troops. An essential feature of the question, What . . . and how? is at a practical level—examining how to actually assess specific conditions and their possible policy significance. Chapters 7 through 9 will attempt to answer this question by exploring the following operational sets of criteria:

- Reasons for government involvement in environmental justice.
- Approaches to measuring environmental socioeconomic impact, whether by governmental or nongovernmental organizations.
- Demonstration of a method for assessing environmental-justice conditions that involve multiple risks.

Chapter 7 begins this exploration: it argues that from the perspective of maximizing overall social welfare, government involvement can be justified only in the presence of very specific socioeconomic conditions. And in the specific case of environmental justice, there are instances where these conditions may not occur. Most of these necessary and sufficient conditions revolve around the occurrence of what economists call *market failure.* Of course, all government intervention does not entail responding to market

101

failure. But it may be argued that in the case of most environmental-justice problems, market failure should be the guiding criterion for any corrective government involvement.[1] Ignoring this guiding principle, government intervention in non-market-failure environmental-justice situations, while perhaps assisting a few, will often result in a worsening of society's overall welfare.[2]

But recognizing the reasons for government involvement in the abstract and identifying specific occurrences in the real world are two very different exercises. As will become clearer in Chapters 8 and 9, the issues of government involvement in many environmental-justice cases greatly depend on proper measurements of environmental-justice conditions and occurrences. In fact, a major stumbling block in developing a comprehensive federal environmental-justice policy is the poor condition of available measurement, or lack of understanding of the measurement, of occurrences of environmental injustice. Poor or inappropriate measurement in this case can lead to misguided intervention decisions. This problem returns us to the axiom that mere observation of an instance of unequal distribution of environmental burdens and benefits does not automatically qualify an event as a case of environmental injustice.

Taking this a step further, and assuming that the question of the form of intervention is resolved, the challenge becomes how to select the appropriate level and scope of government involvement. That is, is the government activity one that should be primarily local or state based, or is it more properly federally based? Or should it be a multilevel endeavor? To add to the complexity of the problem, most area-specific environmental-justice problems are just that: area specific, local in both their source and effect. But does this locality parameter mean that the appropriate level of government involvement should also be local? What if similar local patterns can be detected over a national or regional area? What about non-area-specific problems such as farm worker pesticide exposure, lead poisoning, or contaminated fish consumption? Are state or federal actions the proper response to these problems? What about a purely local system that exhibits locality differences that resemble the prisoner's dilemma? That is, those localities with the more restrictive local rules are placed at a competitive disadvantage with other localities in trying to encourage business location in their community. Will not a local-only solution result, nationally or regionally, in a less than optimal outcome?[3]

Political realities and needs will undoubtedly shape much of the government environmental-justice agenda. Nevertheless, a more careful assessment of the question of government involvement still offers a background or foundation on which the politics of action can be played out. In many cases, it is not a choice of a politically viable but inefficient response versus a politically unpalatable but efficient response. Many policy mistakes in past regulatory actions, especially in the environmental area, have been due less to a concession to political realities and more to simple ignorance of the dynamics of the socioeconomic forces involved. Even politicians occasionally like to make sense when given the opportunity.

Role of Public-Policy Analysis

In trying for a more rational approach to understanding, assessing, and responding to the issue of environmental justice, a natural decision is to turn to a problem-framing and decision-making approach such as public-policy analysis. Granted, no real-world policy activity in the environmental area, or most other areas of government intervention, reflects a purely "rational" response to a socioeconomic problem. Embedded in any policy decision are such factors as the appeasement of constituency groups, political expediency, and many other objectives in which resources are used inefficiently. But while admitting that all choices are not immediately rational, such an admission does not diminish the attraction of evaluating an issue such as environmental justice from a perspective other than pure emotionalism or political expediency.

An important point to recognize about most environmental-justice issues is that they seldom involve purely social policy circumstances.[4] Whether it is decision-making about site location, the use of pesticides in farming, or protocols for the clean-up of hazardous sites, market-driven forces play the major role in outcomes, regardless of other prejudices or biases. There is no evidence that firms or business enterprises, whatever the biases of their decision-makers, choose, for example, minority communities as the sites for a particular burdensome environmental activity out of some desire to discriminate against minorities. Firms, like people, seldom discriminate unless it is in their best perceived interest. Minority communities often lack political clout, lack risk information, are in a weak economic position, or have other disadvantages that make them, purely from a market-driven perspective, a logical location choice. Thus, policy solutions that focus on changing underlying biases and attitudes while ignoring the market realities that allowed these biases and attitudes to reach fruition seldom succeed. And hence, viewing environmental-justice issues through the lens of public-policy analysis is not just a curious exercise, but a necessary condition for successful government response.

The field of public-policy analysis over the last twenty to twenty-five years has rapidly developed as an interrelated collection of methods, concepts, and constructs for evaluating, understanding, and guiding government intervention into what previously have been essentially private market activities. Borrowing heavily from economics, management science, and to a lesser extent political science, public-policy analysis attempts to provide a guideline for possible government response to socioeconomic problems, and at the same time, it provides a framework for assessment of the subsequently developed public activities. Although far from perfect, the policy-analytic approach definitely offers a more structured and rational basis for what to do and when to do it. To the problems at hand, a policy analysis approach to environmental-justice issues does not guarantee solutions, but it does offer useful insights into where to begin looking for them.

Seven

Environmental Justice through the Lens of Policy Analysis

Why Should Government Get Involved?

It would be naive to believe that any real-world policy response to the challenge of environmental justice can perfectly follow the principles of policy analysis. Conversely, a policy response that completely ignores these principles invites disaster.

Strategies for dealing with the environmental-justice problems of this country have ranged from a complete denial of their existence to a call for massive and expensive government intervention. In several of the earlier chapters, I addressed the first strategy by arguing that proving particular groups have endured unfair environmental burdens, or have been virtually excluded from this country's environmental policy-making process, is not the primary issue of environmental justice. Far more important is recognizing that the socioeconomic and risk effects of environmental policies and activities are many and varied.[1] As for the second, diametrically opposite, strategy, the mere existence of unequal distribution of environmental risks or benefits does not automatically justify government involvement, much less a massive and expensive effort.

A Basic Principle of Public-Policy Analysis

"The creation of public policy should be, as much as possible, a rational process." Admittedly an ideal, this declaration means that actions and policies are evaluated by some specified criteria.[2] In the same way, public-policy analysis views government involvement in society as a remedy to be sought only when private solutions do not work.[3] More important, from a policy-analytic perspective, even when private solutions fail, government should intervene only if it can improve the overall situation.

According to public-policy analysis, pork-barreling is insufficient justification for public action. Furthermore, when a justifiable action is taken, the objective should be achieved with the most efficient use of resources and should make the socioeconomic environment better than it was before. But policy analysis goes beyond simply identifying when government intervention should occur; it also suggests the nature of that intervention. For example, regulations preventing or significantly reducing certain air emissions from industrial sources have been recognized as a public responsibility. But from the perspective of policy analysis, the often-employed command-and-control regulatory approach has proven less than optimal. In this familiar method of environmental activity control, compliance involves the firm satisfying a set of detailed regulations for equipment and procedures. Command and control, under many circumstances, fails the criterion of efficient use of resources: in many cases, companies end up spending more than they need to on air emissions control, and the public fares no better, and in some cases worse.

When policy actions address issues of environmental justice, the temptation to jump right in and begin "correcting" problems is great.[4] Great injustices appear to be occurring and seem to cry out for an immediate response. Furthermore, no one wants to believe that their inaction resulted in unnecessary suffering or even death. The urgency of some environmental-justice problems does not diminish the need for a systematic assessment of both the situational conditions and the available policy alternatives. Following ill-conceived policy remedies in the short run can have disastrous consequences in the long run. For example, it is frequently proposed that regulations be passed to ensure that all environmental burdens are equally distributed among the population. However, the idea of enforcing equal distribution of environmental burden is ill-advised public policy; if achieved, it would actually mean an inequity in social welfare.[5]

Public-policy analysis is not simply a way of justifying public activity, such as regulations or the public-sector provision of some services; it also provides us with some clear criteria for evaluating the consequences of various public actions and policies. In the case of environmental justice, which is not a single condition but a wide range of problems clustered under one umbrella, the principles of public-policy analysis should serve as beacons that guide the current development of policy.

Why Do Governments Get Involved?

Modern economics—and by extension, its progeny, public-policy analysis —begins with the ideal of a perfectly competitive economic environment, in which production and consumption of all goods and services is perfectly efficient. At its simplest, this concept means that no resource is used in an industry or at a level of production when it could be used in another industry or at another level of production and thereby result in someone being better off—without harming the welfare of someone else.[6] Efficiency in such a finely balanced system also means that the mix of consumption— both in terms of level (i.e., how much you are producing) and type of goods and services—cannot be changed to make someone better off without at the same time making someone else worse off. With minimal government interference, prices and quantities are determined by the force of many individuals, each with his or her own unique preferences, and each pursuing his or her own utility-maximizing objectives. Aggregate economic welfare thus achieves higher levels than under any other possible organizing structure.

If the above conditions actually did occur, does this mean that the only role of government would be to maintain and enforce the underlying rules of conduct in market exchanges? Of course not. Pursuing efficient markets is not the only possible societal ideal. In fact, it may not even be the most important. For example, efficient markets could occur under social and political conditions that we as a society would find repugnant. Equality, human dignity and rights, and political participation are among the societal goals that have at times superseded the goal of economic efficiency. Society may knowingly sacrifice economic efficiency in order to secure a more equitable distribution of economic resources or to improve the welfare conditions of some members of society.[7]

Of course, except for some special small, isolated, and somewhat artificial environments, a truly efficient economic environment does not exist. Thus, in many cases, there need not be a trade-off between economic efficiency and equity or other societal values. Efficiency and equity can be advanced simultaneously. Furthermore, modern public-policy analysis finds that even when pursuing other societal goals, such as increasing socioeconomic equity by using the power of markets and utility incentives for individuals—although it is not a holy grail—the societal goal can often be achieved more quickly and with the expenditure of fewer resources.

In the specific case of environmental justice, much time and energy have been devoted to arguing for solutions to particular problems on the grounds of equality, equity, or human dignity. But even a cursory examination of many of these solutions reveals an ignorance of the economic and social forces that produced the particular problem. Even worse, given their lack of appreciation of market dynamics, many of these solutions are either guaranteed to fail or, at best, to produce a small gain at a huge price. In contrast, a policy-analytic approach to environmental justice devotes as much attention to understanding the how and why of a particular issue as it does to its results. For example, subsistence fishing in the American Great Lakes poses a threat of toxicity to certain minority populations, and the practice has been labeled an environmental-justice issue. But no one is forcing these individuals to fish; they are motivated instead by economic need and a lack of viable alternatives. A simple ban on fishing will not relieve the economic forces that prompt the fishing. And although stricter regional regulation of pollution may offer some long-term relief, such an approach does not change the immediate dynamics of the underlying subsistence need. A policy-analytic approach would recognize the futility of simply banning fishing, except as a convenient political gesture. And at the same time it would recognize that imposing several million dollars in annual pollution control costs just to save several thousand dollars in individual food costs is a poor economic trade-off. A wiser policy-analytic move would focus more attention in the long run on developing economic incentives to redirect the behavior of the individuals involved and on disseminating clearer information on the consequences of continuing to eat contaminated fish. At the same time, in the short run, this approach would consider food subsidies for the target population as a far cheaper alternative.

But does adopting a strong policy-analytic approach with an emphasis on market efficiency mean that governments should restrict themselves to "purely" political activities and stay away from market regulation? Absolutely not. There are still strong reasons for government involvement in economic activities, even while pursuing the ideal of efficient markets. All these reasons revolve around the phenomenon of market failure. In other words, under a number of conditions, a purely private market solution to a market problem will not result in efficient or fair use of resources. In these instances, allowing the market to operate without government intervention results in both volume and pricing of resources, goods, or services that are either too high or too low.[8] It also can result in an inequitable (different from an unequal) distribution of these benefits and costs.[9] Finally, and possibly most significant, market failure can result in a worsening of the condition of society as a whole.

Consider the case of a hazardous-material treatment facility whose operation poses a serious health risk to the surrounding community. A private exchange between the facility operator and a company wishing to dispose of some hazardous materials does not include the risks to health and safety, along with their associated costs, that the community must endure. In this case, the community is essentially providing an involuntary financial and health subsidy to the facility operator.[10] The operator is not paying the true cost—what economists call the social cost—of production and will tend to operate at a higher level of activity than is economically efficient, and the company will pay a lower price for the disposing service it needs than is resource efficient. The facility operator and the facility user have both received an unfair and unearned benefit. Policy analysis shows that such a *negative externality* results as a spillover from the private transaction in which the community is an unwilling partner. This is market failure.

Charles L. Shultze identifies "four sets of factors whose existence leads to market failure, and also limits the range of corrective action available to society":[11]

Transaction costs and externalities.
High information costs.
High degree of uncertainty.
The "free rider" problem.

Further complicating the matter, in most cases of market failure, more than one of these factors will be present. For example, not only might externalities exist in many site location problems, but considerable uncertainty exists about the degree of joint risks the community faces, and whatever information can be obtained has an extremely high cost. In forming a particular environmental-justice policy, it is important to identify each factor contributing to market failure. Policies that focus on only one or two causes of market failure but ignore others could produce either weak or counterproductive solutions. A solution to an externality problem that assumes knowledge about impacts that cannot easily be obtained is almost worse than no solution at all. Before considering each of these four factors and how they relate to specific environmental-justice problems, a word of explanation. All market exchanges involve, to varying degrees, transaction costs, uncertainty, information costs, and even a "free rider" problem; it is simply a matter of degree. Market failure occurs at the point at which one or more of these factors becomes so significant that the private market results fall far short of an efficient solution. Of course, the definition of *significant* is entirely subjective; one could spend considerable time debating the exact threshold of market failure. This judgment, for now, I do not make.

High Transaction Costs

High transaction costs often top the list as the most compelling source of market failure. From brokers' fees to telephone charges to the time and energy spent searching for a new house or apartment, all market exchanges involve costs beyond the "selling price" of a good or service. But some markets involve side effects, or externalities, that even when recognized may be extremely difficult or costly to incorporate into the exchange process. In public finance or economics, this is also part of what is called *non-appropriability*—a situation in which the provider of a service cannot effectively receive all returns from a transaction, or be held accountable for all costs.

For example, several transaction problems occur with the location of a hazardous-material facility in a community. First, assume that such a facility would increase the health risk to the community, or at least cause the diminution of community property values. If the market exchange is limited to the operator and facility customers, then the community has provided an involuntary subsidy to the operator. That is, one cost of the transaction—increased health risk—is borne by neither the facility's operator nor its customers. Rather, the community bears the costs but receives no corresponding payment.

Even if the government creates a new "property right"—which is itself a form of government involvement—for some degree of environmental risk protection for the community, that action still does not necessarily satisfy the preference of each member of the community. This is not a three-way exchange but a thousands-way exchange. The costs of obtaining and acting on each individual's preferences for health and perceived risk prohibit a simple market solution.

Government involvement in such a case may take as simple a path as recognizing the community's right to at least some protection for health and property value. However, this level of involvement would provide no specification on how to guard or enforce these rights. Conversely, government could, in recognition of the difficulty of transaction costs, adopt the severe stance that no action is permitted that lowers either community health or property value. Between these two extremes, a more likely approach would be for government to specify a mechanism for collective community representation, or to avoid the property rights issue by setting some standards of risk that the company has to follow.

An even worse scenario is encountered in the environmental-justice problem presented by the cumulative risk of a community's exposure to multiple sources of risk. In this situation, failure due to high transaction costs also plays a major role. The transaction costs for negotiating a private market solution between the community and a diverse group of firms would be enormous. Each firm would have a different responsibility, such as how long their facilities have been at the site and the degree of risk their production represents versus other firms in the mix. Each community member would have equivalent dissimilarities of risk preference and exposure. A private market solution would be almost impossible.

Government intervention in this case can take several forms. The high transaction costs may be reduced via regulation of the negotiation process. Or the cost may be internalized by direct regulation of the various operations themselves (emission limits, etc.). There may be other government responses to high transaction costs, but the critical guiding principle is that government intervention focuses on the cost of negotiating the problem. It does not directly dictate behavior or try to restructure the goals of either firm or community. The individual players are left with the right to respond in the way they feel is best for them, within an exchange environment that has been shaped by the government.

High Information Costs

Especially troublesome in resolving some existing environmental-justice problems, and in getting cooperation in avoiding or mitigating future problems, is market failure due to high information costs. One of the basic principles of efficient markets is that everyone involved in a transaction operates with full knowledge of all its costs and benefits. Even if such information is less than perfect, the market process still works well when all parties operate from the same general level of access to and understanding of information.

Unfortunately, in many environmental-justice situations, perhaps the major information problem occurs when one party comes to the market exchange with significantly less information than the others. Consider our example of a siting decision that exposes the surrounding community to health risks. The facility's operator, one of the major players in the exchange, usually has a vastly greater and more realistic understanding of the facility's impact on the community. The individual community members, all of whom have different needs for this information, along with different degrees of potential risk, would seldom have the resources to acquire, either individually or collectively, the necessary information about costs and benefits that would help level the playing field.

Ignoring the issue of dollar costs, how efficient is a market exchange process that, in order to work, requires individuals to acquire risk information that they have neither the training nor the time to properly evaluate? Interpreting the information affecting siting decisions requires the skills of highly trained technicians and analysts. Are all community members thus expected to verse themselves in environmental chemistry, facilities operation procedures, and risk assessment before any exchange?

But isn't the problem of asymmetry of information overstated? Doesn't there currently exist a nonregulatory solution that would prevent market failure? If health problems do develop as a result of exposure to the facility, affected community members should be able to collect damage payments via the liability laws. Would these liability statutes, which were developed partly to address the one-sidedness of information during market exchanges, eliminate much of this inefficiency in market exchange? Realizing their liability for unexplained hazards, wouldn't the more informed party have an incentive to provide to all less-informed participants as much information as possible in order to avoid future liability damages?

The answer to all these questions is no. The problem is again one of information. Environmental health risk, with the exception of specific accidents, is seldom immediately self-evident. Even if it can be shown that a community suffers poorer health than other communities, specific party assignment of liability would be nearly impossible. Too many other factors, such as the lifestyles and histories of the individual community members, may be cited as concomitant contributors to compromised health. Even assuming the extraordinary—that reliable risk information could be obtained by the community itself (for example, conclusively showing that the facility's operation increased the incidence of skin cancer among citizens by a significant 30 percent)—current liability law in most states does not provide for partial assessments. Either liability exists or it does not. As a result, operators of hazardous facilities have considerably less motivation to engage in the type of information-sharing suggested above.

Add to the above scenario the fact that individual community members do not share the same degree of exposure, either by length of exposure time or by geographic location, and the problem gets worse. Individuals constantly enter and exit the community rolls. Imagine the information costs for keeping track of every individual's exposure history. The information cost for a single facility could easily exceed any compensating payments.

The high cost of information also confronts some farm workers exposed to toxic pesticides. Hispanic farm workers, particularly migrant farm workers, have neither the financial resources to obtain necessary information on the toxicity of the myriad chemicals they are exposed to, nor, in many cases, the expertise to interpret that information if they did have it. And the linkage of exposure to liability here is not much better than in the siting of hazardous facilities in a community. Most of such farm workers are temporary employees who move from one farm to another. Among such a migrant population, the problem of compound and concomitant effects is enormous. Furthermore, other than the obvious and immediate effects of exposure to acutely toxic chemicals, most pesticides have a cumulative effect that becomes evident only in the long term—as much as several years after the initial exposure. Thus, individual farmers have no immediate incentive to reveal or to protect against such effects. This is also market failure.

To deal with the problem of high information costs, the government could delegate the responsibility of sharing information to the more informed parties in a transaction. Alternatively, the government could directly provide and interpret information. Also, rather than deal with the information problem, the government could regulate the risk or process itself. This last approach is part of the command-and-control approach to safety regulation so favored by the Occupational Safety and Health Administration (OSHA) and the U.S. Environmental Protection Agency. As noted earlier, while the command-and-control approach has many drawbacks, in some cases, this approach may be the only alternative.

High Degree of Uncertainty

Uncertainty is the direct result of inadequate information. In some cases, the cost to correct this information inadequacy, and thus to reduce uncertainty, is high; in other cases, the information is simply unavailable at any price. Both high costs and a lack of availability of information contribute greatly to uncertainty in the area of environmental justice. When one party to an exchange is well informed—usually a firm—and the other is poorly informed—usually the individual or community—the resulting uncertainty on the part of the uninformed adds to the inefficiency and inequity of outcomes.

But firms can also experience uncertainty, as when an existing facility appears to be damaging the health of a community. If the firm volunteers to assume some responsibility for the damage, it may now be exposed to considerable liability for all subsequently detected health effects—for which, in all likelihood, it is not wholly responsible. The rational strategy of a firm faced with such uncertainty of future liability is to avoid admission of any liability at all costs. Such a strategy in turn means a suboptimal outcome with a retardation in the process of improving the quality of the environment.

High uncertainty contributes in another way to the inability to take action. If no existing methodology exists for assessing the actual consequences of joint exposure, neither conclusive remediation nor compensation can occur. And when the consequences cannot be isolated or quantified in any reasonable manner, it is not even possible to create property rights to those effects. That is, assigning rights to a certain level of risk protection is very difficult when the level of risk itself is unknown.

Government can take several different routes in dealing with high levels of uncertainty. When long-term liability and outcomes are uncertain, the government can either release some parties from liability in order to encourage immediate action, or it can assume some of the future liability or future investment itself. The latter route was chosen in the case of the liability associated with the swine flu vaccine of the 1970s. The levels of uncertainty and risk were high enough that no single private party was ready, at any level of compensation, to assume responsibility for the undertakings. In the same manner, in the 1950s, 1960s, and 1970s, without government assuming the extraordinary capital risk and huge initial information and research costs, the space exploration program would have been impossible.[12]

A major problem of environmental policy today is posed by the reclamation of "brownfields"—the return of formerly contaminated sites to viable community use. The EPA has actually developed a Brownfields Economic Development Initiative to devise methods for reclaiming such currently useless and economically draining sites.[13] The significant dilemma facing the EPA initiative and several state brownfields programs, such as those in Michigan and Pennsylvania, is not that they must improve technical knowledge about the toxic effects of the sites, but rather that they need to develop procedural methods for reducing the uncertainty of liability associated with the sites. Under present laws, this uncertainty can extend not just to the original owners of a site but also to lending associations that may have a financial stake in the site, future owners, and even to past minor, but financially well-off, users of the site.[14] Uncertainty has paralyzed all players and has essentially placed such sites in a legal and developmental limbo.[15]

The government initiatives in the case of brownfields—which by extension may be applicable to several environmental-justice cases—directly address the problem of uncertainty of liability. Suggested remedies have included offers of "covenants of no sue," by which future owners are relieved of some part of the liability for past contaminations. In some proposals, lending institutions—which in some cases have been too frightened to foreclose on properties out of fear of assuming a liability far in excess of the value of the property—have been released from liability. Yet another provision imposes a government system that tracks the complete history of a property to avoid inappropriate future use.

The Free Rider Problem

The free rider problem is really the problem of a public good.[16] In such a case, one party cannot capture all the benefits of or be held accountable for all the costs of a good or outcome that is to be enjoyed by many. Consider an activity, such as a toxic treatment facility, with potentially serious negative externalities. If individuals in a community negotiate with or pay a facility to modify its operation, all in the community benefit, whether or not they participated in or paid for that modification. A wholly voluntary arrangement may simply have too many incentives for individuals to be free riders. That is, if someone else in the community pays the price, you take a free ride.

From the producer's side, the exposure to agricultural pesticides and the siting of hazardous-material facilities are also free rider problems. Individual agents realize that although they provide information or corrective adjustments, not all the benefits will return to them. Other farmers or site operators who did not contribute to improving conditions would still enjoy the benefits of a healthier community or worker population. It follows that no individual facility owner or farmer has much incentive to personally assume these costs to produce a jointly enjoyed benefit.

Without government intervention, the protection or production resulting from such a free rider public good will be below the economically efficient optimal level. In this case, the optimal level will always be higher than the level of a purely private solution.

The Federal System as a Problem

Although it is not one of the classical reasons for market failure, the federal system of government, in which both state and federal authorities often monitor the same business activities, can also lead to market failure. In the environmental-justice area, this can take several forms. The foremost of these problems occurs when responsibility for resolving specific differences in environmental impacts in a given locale are left exclusively in the hands of local or state government. As the result of such a strategy, a different set of enforcement conditions would prevail in each state, or even in each locale. In some states, communities may possess virtual veto power in negotiating with economic entities operating in their midst, while in another state, the decision and rights to locate or operate may remain strictly in the hands of the landowners and the incoming firm.

In a system of fragmented environmental regulation, firms, being rational, profit-maximizing entities, naturally seek out those states that will impose on them the lightest regulatory burden. Currently, companies engage in such "shopping" as they make their investment decisions, and there is no reason to expect them to adopt a different pattern when confronted with growing environmental-justice legislation.

When environmental regulation is fragmented between federal and state governments, the problem is less whether government should intervene than what level of intervention is necessary. Remember, most area-specific environmental-justice problems, such as site location decisions and the remediation of existing sites, are really local in nature. The most efficient resolution to such a problem usually requires a local interpretation, not the imposition of a one-size-fits-all federal rule book. Yet as noted above, major differences in enforcement or standards of compliance directly lead to a patchwork of uneven and inequitable outcomes. There is no simple answer to this dilemma, but one alternative federal government role may be to establish a standard of regulation below which no state may venture. To establish regulations much past such a baseline would require micromanagement, at which the federal government has shown itself to be particularly inept. Conversely, to allow states to take the lead in setting their own standards for environmental justice or social impact protection invites regional inequity.

New Rights

In our modern society, the rights of individuals and groups, especially property rights, are constantly evolving and being redefined.[17] We say that an individual or a community has a right to a certain level of clean air, or that an industry has the right to do whatever it wishes as long as it keeps its harmful emissions into air or water below a certain level. This process of rights creation and assignment reflects not so much greater understanding of business activities as it does an evolving concept of what constitutes basic human rights. At one time (and in some countries, still), humans did not have the right to avoid being considered property themselves. Put another way, at one time, people had the right to own other humans as property, and that right has finally been taken away. But in most modern cases, rights redefinition has involved the creation and assignment of new property rights.

The creation or recognition of rights is a dynamic process that responds to constantly changing socioeconomic and political realities.[18] Nowhere has the creation and assignment of property rights been more prevalent than in the area of the environment and natural resources. Before the 1960s, business treated air and water as essentially free goods. Today, a minimum standard of air and water quality is taken for granted as the right of every citizen. In fact, in some cases, as with human property, citizens cannot even give away or sell these rights.

Environmental justice—the idea that all groups of people have a fundamental right to enjoy equitable levels of environmental costs and benefits— is an emerging concept. This right to equitable burdens and benefits is not part of the original mandate of the environmental movement or of the civil rights movement. As noted elsewhere, all policy issues evolve at different times in response to different conditions. Today, the concept of environmental justice is on the verge of being incorporated as a "new" right. The eventual form and nature of this right is still not clear. But from a policy-analytic perspective, as this right emerges, it must be incorporated into the policy assessment process.

As the policy agenda for environmental justice becomes more clearly defined, the following questions must also be addressed: What are the property rights of a community or individual? What are the property rights of a firm engaged in economic activity in a community? These are not so much questions of economic efficiency as of social organization.

Where business activities impinge on community life or where there are incompatible uses for the same commonly held resource, such as air, the creation and assignment of property rights is a major concern. Careful assignment of property rights can reduce much of the potential ambiguity and the resulting resource inefficiency associated with such environmental and natural resource issues as air and water pollution. Even where assignment of property rights does not resolve the issue directly, it will very often provide clearer grounds for its subsequent resolution.

From the perspective of classical economic efficiency, precisely who receives which rights may be irrelevant.[19] Whether a community or a firm holds the property rights to a resource or its use, the resources can nevertheless be used efficiently.[20] But from the standpoint of equity distribution, the assignment of rights makes a big difference. If a firm is held blameless for any negative effects on a community, in theory, that community could negotiate with the firm until an efficient outcome is achieved. In practice, however, this is simply not a viable option for many poor communities. If a firm has the property right to pollute, many poor residents must either accept the risks or leave the community. Conversely, assigning resource use—for example, property rights to a community—forces the more resource-rich firm to initiate, and follow through with, negotiations.

What Is the "Right" Action?

This chapter has dealt with the reasons why government may become involved in the solution of environmental-justice problems. But beyond a few comments about the inefficiency of direct, command-and-control regulation, I still have not explained what, if anything, government should do. In the next two chapters, I will further address this question.

Eight

The Measurement of
Environmental Justice
Some Rules of Engagement

By involving demographic, geographic, and environmental activity information, the investigations and analyses of a large class of problems in environmental justice require developed assessment data. Current environmental data—which reflect, among other things, a priority system that devalues concerns about social impact—are woefully inadequate to provide a clear assessment of any but the most egregious cases of harmful social impact. Current methodologies for assessing risk would have to be radically altered before we can or will properly address such problems as the location of hazardous-waste facilities or the disproportionate effects to human health of pesticide use. Moreover, misapplication and misinterpretation of what information is available muddle the discussion about environmental justice.

From the perspective of policy analysis, government intervention in an issue area such as environmental justice is most often prompted either by market failure or an inequity in the distribution of environmental benefit or cost.[1] Such an unequal apportionment can occur even when no perceivable market failure has occurred. But before government can correct market failure or distributional inequity, the phenomenon must first be identified systematically. It is not enough that individuals *feel* disadvantaged or even that some governmental or nongovernmental organization declares that a particular group has been victimized.

In the heated debate about whether an environmental-justice problem

118

exists at the national level, critics on both sides wheel out their analytic or emotional guns and fire away at the "opposition." Critics of the environmental-justice argument cite flaws in the admittedly weak data and methodological approaches of some of the earliest environmental-justice studies.[2] Those claiming the existence of an environmental-justice problem criticize the narrow focus on race by some of the critics of environmental justice. These supporters argue that their opponents miss the broader message of the movement, that in the United States environmental activities and policies are insensitive to the welfare of major segments of its population. To bolster their arguments, these supporters then produce their latest study that shows that in a particular region, even when using the smaller measurement units or a different statistical methodology, racial inequity in environmental impact can still be observed.

For better or for worse (mainly worse), measurement—and controversies over measurement—drives a large part of the current discussion on environmental justice. There are many good questions and few good answers. For example, what is the right geographic level of assessment? What is the right set of statistical or analytic algorithms to employ? Which indicator variables should be included in an environmental-justice study, and which are a waste of time and resources? Here begins a technical section of the book, in which I investigate several aspects of the problem of measuring, evaluating, and making policy recommendations on environmental justice. In the present chapter, I describe a set of principles with which to approach the identification and measurement of environmental-justice problems. It is important that these principles not be viewed as a precise and exhaustive prescription for measuring the many different types of environmental inequity. Only a broad outline is drawn here; the details depend on individual circumstances.

Of course, some readers who live in communities in which toxic waste and other extraordinary environmental burdens are an ever-present threat may feel that instances of environmental injustice are obvious and that we should move right to the consideration of remedies. And another large group, one more interested in exploring the ethical and moral aspects of environmental justice, may be overwhelmed by the quantitative and analytic orientation of this entire section. In response, it must be noted that problems of environmental justice are not always obvious, and most of these issues in specific cases cannot be appreciated without some understanding of the theoretical and practical problems associated with measuring and evaluating the phenomenon.

The following story illustrates the first (and in some ways most important) lesson about the measurement and evaluation of environmental justice. While I was visiting the U.S. Environmental Protection Agency to research the topic of environmental justice, a representative from an environmental industry association paid several calls on the agency's Office of Environmental Justice. The representative had one major mission. The leadership of the association was under pressure from its members to determine how environmental justice was currently being, or was going to be, measured. Members wanted to know what method or methods they could employ to examine possible problems of environmental justice in their industry. The representative was extremely disappointed when told that no single method for assessing environmental justice existed, or was ever likely to exist.

This, then, is the first lesson of environmental-justice measurement: there is no single method or collection of methods for directly measuring any environmental-justice problem. Because environmental justice covers such a wide range of situations and conditions, no single methodology, however powerful and however complicated, can ever successfully assess it. Even when confronting what appears to be a single-dimension situation, such as site location bias, there are many different aspects of the problem, each requiring different assessment mechanics.

What can be measured are outcomes or conditions that indicate the possible presence of a particular type of environmental-justice problem. What can be measured are differences between groups of people with respect to an indicator such as mortality rate, proximity to an identified environmental hazard, or incidence of cancer. What can be evaluated are the possible problems in a decision-making process that penalize a particular group of people. Whether they are static (evaluating at a single point in time) or dynamic (spanning a time period), individual methods capture only pieces of the puzzle. No matter how many current methodologies are employed, the ultimate decision about whether an environmental-justice problem exists remains one of interpretation.

Furthermore, the methods for assessing the existence or outcomes of an environmental-justice problem vary with the circumstances. Sometimes it may require extremely sophisticated, mathematically involved techniques just to detect a problem. At other times, all that may be required is a simple observation of the manner in which a decision is being made. It is extremely unlikely, however, that a foolproof method can ever be developed for conclusively showing that a given condition is an environmental-justice problem. There will always be concomitant factors and missing data to confound the issue. The ultimate decision about the existence or causes of an environmental-justice problem—whether settled by courts, regulatory mandate, or arbitrated agreement—is a value judgment dependent as much on nonquantitative evaluation as on quantitative analysis.

Since 1993, and perhaps earlier, several units at the EPA have been trying to devise systems for measuring and assessing conditions of environmental justice. This same exercise is simultaneously being pursued by a number of nongovernmental organizations on both sides of the environmental-justice debate. A common note heard from many of these questers is the need for a single method they can agree on for performing this task. Unfortunately, as noted earlier, there can be no agreement because environmental justice is not a single, simple phenomenon, and thus no single methodology is applicable in all situations.

Environmental justice encompasses a whole array of issues and phenomena involving the differentiated social impact of practices and policies, both public and private, on the environment and natural resources. Evaluating these many manifestations must involve an array of methodologies and constructs from both the social and physical sciences. The best that can be accomplished at this time is a classification of problem types and an identification of the best available approaches for assessing each type.

Whether one is a program officer at a regional or state environmental affairs office, an investigative agent for a state economic development agency, or a policy analyst for an environmental-justice advocacy group, some common operational definitions and guidelines may help in deciding when an environmental-justice problem exists and in assessing the extent of the problem. In proceeding through this exploration, remember that many of the definitions and subject divisions are arbitrary. A guiding principle is that the classifications as such make both analytical and operational sense. That is, will these classifications work in real-world settings? The watchword here is flexibility. There is nothing sacred or absolute about the following forms. If other terms or forms make more operational sense, develop them. This is just a starting point.

There are a number of problem dimensions that should be determined at the outset of any analysis. But an accurate assessment of these dimensions is less important than their careful classification. As in the analysis of many policy problems in the real world, where quantitative (and even qualitative) measurements are often nearly impossible to obtain, the mere exercise of classification can still give the analyst greater insight and often provide the affected parties with clearer spheres of discussion.[3]

Major Dimensions of an
Environmental-Justice Problem

In the first stage of analysis, it is critical to understand that many situations of environmental justice are composites of more than one type of problem. Thus, more than one assessment method or solution may apply. It is essential in any assessment to recognize such compound problems and to separate, as much as possible, the conditions into their individual components. Failure to properly separate results confounds an assessment, possibly yielding misleading conclusions. For example, a geographically specific problem such as high exposure to environmental hazards in a given neighborhood may require policy decisions in two different time dimensions, present and historic. It is quite possible that many of current residents were attracted to a neighborhood because preexisting environmental hazards lowered the cost of housing. Nevertheless, these same residents may not wish to increase the extent of environmental hazards in their community. It may be argued that the historic problem is reflected to some degree in the existing housing values, and because individuals entered the community with an awareness of the potential risks, little remedial action is called for.[4] On the other hand, the introduction of a new hazardous site to the community—a decision that affects the present—is a much more serious problem. The present siting problem may require intervention, may involve conflict resolution, and may involve issues of rights assignment and community representation in an active decision process. The key dimensions are problem type, placement in time, and measurement dimensions.

Problem Type

In its short history, environmental justice has tended to focus on area-specific exposure to hazardous materials. But although this category accounts for a significant number of recognized environmental-justice problems, it does not account for them all. The first step in assessing an environmental-justice problem is to determine which type it belongs to. Because methods of assessment differ considerably between the various types, careful classification of a problem will save wasted effort later.

GEOGRAPHICALLY SPECIFIC PROBLEMS

Geographically specific problems occur when a specific neighborhood or set of communities either suffers disproportionate burdens or fails to receive proportionate benefits. These geographically specific problems can be further subdivided into those with a specific population bias and those without. Included in the former subtype would be African American or low-income communities with a disproportionate share of hazardous materials sites. The latter subtype would include situations with just regional or local biases in the exposure to environmental hazards. In this latter case, it is not race, class, or income that determines the inequity, but simply where one lives. For instance, the location of a hazardous-waste site in a politically weak or disorganized rural area could be an environmental-justice problem that may not correlate with race, income, or education.

With rare exceptions, geographically specific problems will involve some measurement based on spatial units. That is, assume that with a geographically specific problem a county, state/regional, census tract, or similar spatial unit will be used to measure characteristics such as site location, income, or health conditions. In fact, in many cases, the same environmental-justice problem may involve measurement at several different spatial-unit levels. For example, in defining and then evaluating a potential neighborhood environmental-justice problem, both census tract and some larger spatial unit may be used in evaluation.

This point about multiple spatial-unit measurement deserves special note. The familiar debate in many geographically specific environmental-justice studies has been over the proper spatial unit of measurement. Is it to be census block groups or census tracts or larger still—wards, postal zones, or even counties? This debate has largely been a waste of time. Usually in any practical problem situation, the proper analytic protocol would be to use several different spatial measurement units and to be sensitive if or when the indications of environmental-justice problems change. Investigating the shift points provides potentially crucial information on the nature of the environmental-justice problem—information that is completely lost if the analysis is performed at only one spatial level. Worry less about which unit to use, A versus B; just use them both and see what happens.

POPULATION-SPECIFIC PROBLEMS

Population-specific problems occur when the life experience or behavior of a group results in their inequitable treatment. Examples include pesticide exposure of Hispanic farm workers; exposure to toxins through subsistence fishing by low-income populations fishing in polluted waters; the significantly higher fish consumption by several Asian populations regardless of income, accompanied by the exceeding of acceptable toxic exposure levels; and greater exposure of minority or low-income individuals to lead poisoning from the lead paint in older and poorly maintained housing.

The important characteristics of such population-specific problems are, first, that they tend to be nongeographically specific, and second, they often involve actions or conditions that reflect some lifestyle characteristics of a population. For example, because the poor are more likely to reside in poorly maintained inner-city residences, they will more likely be exposed to higher levels of lead paint poisoning. In the same manner, because the poor in some Great Lakes areas and other locations near water have for long periods relied on subsistence fishing, they now face a greater risk of toxic exposure. The risk reflects a dietary preference and a reliance on a subsistence resource.

What makes this category problematic are that the forces creating the environmental-justice conditions are less likely to reflect a deliberate policy bias. More likely they reflect choices made by population members themselves. Thus, the question always comes up: to what extent can any policy response ameliorate such problems or, more important, have a right to even address such problems?

ECONOMICALLY SPECIFIC PROBLEMS

In some ways, all problems of environmental justice have an economic component. In this category, however, specific economic needs of an aggregate population define the problem. Examples include Native American communities choosing, under economic duress, to permit nuclear storage on tribal land. Similarly, minority or low-income populations do not have the economic resources to detect and alleviate radon threats.

It could be argued that little distinguishes this category from the previous population-specific one; and at several points, the two certainly overlap. For example, the higher pesticide exposure of Hispanic farm workers is both a population-specific problem and one of economic need. However, from an operational perspective, this category of environmental-justice problem can best stand by itself because problems based on economic need often lend themselves to economic solutions. Furthermore, there is no question of a difference in environmental impact or risk. Economic need often drives deliberate population decisions to trade increased environmental risk for economic compensation. Finally, these problems may or may not be geographically specific.

Where they are economically specific, the assessment, unlike the earlier category of geographically specific problem, may not involve use of spatial-unit measurements. For problems based strictly on economic need, the proper focus of environmental-justice assessment would be on the decision process rather than on the attempt to detect risk exposure differences between the affected population and others. The important questions become whether the population had adequate information about the risks they chose to endure and how much freedom of choice the population actually had in making that choice.

Placement in Time

The second most important aspect of an environmental-justice problem is its placement in time. In fact, environmental-justice problems may be described as having one of three time characteristics: historic, present, and future.

HISTORIC

Some environmental injustice results from historic decisions. For example, past site location decisions endanger present communities or, because of cumulative past exposure to pesticides, farm workers currently endure poorer health. The key characteristic in historic-dimension cases is that most of the risk-creating events have already occurred. Furthermore, in historic problems, while the risk effects may still be felt, the risk-generating conditions or activities no longer operate.

The classic historic problem would be an abandoned steel mill. The mill represents a historic operation that no longer generates new pollutants but does impose potential current risks. In fact, most national priority list Superfund sites represent historic problems.

PRESENT

A present problem involves a current and ongoing condition. For example, while the use of lead paint in older houses represents a historic problem, the present lack of proper preventive maintenance imperils the current residents of those houses. In the same manner, the historic patterns of the operation of a manufacturing plant may have resulted in a hazardous materials site that remains a threat to present community members.

Furthermore, a present environmental-justice problem would involve not the existence of historic sites, nor just the present hazard they pose, but, more important, the differences in remediation among these sites based on factors such as race, income, or ethnicity. In the clean-up of a hazardous site, for instance, some populations may receive more favorable treatment than others. Thus, the present-time assessment may focus more on comparative differences between neighborhoods than on evaluation of the risks encountered in a single neighborhood.

FUTURE

This category includes possible future decisions on site locations, as well as the future behavior of a community or population—that is, decisions on the possible future location of hazardous-material processing facilities, decisions on methods for involving community- or population-specific groups in environmental policy-making, and the future methods of disseminating environmental risk information.

Measurement Dimensions

Regardless of which of the three areas any problem of environmental justice represents—geographically specific, population specific, or economically specific—the same set of contributing factors or dimensions should be reviewed in its evaluation. A systematic assessment of each of the following factors provides an essential framework on which to build further analysis.

Demographic Factors

Demographic factors are the dimensions across which comparisons can be conducted to ascertain significant outcomes or conditions differences. Of course, not all these demographic factors may operate in a given situation. Nor do observed differences between demographic categories automatically imply an environmental-justice problem. Nevertheless, such differences should alert the policy analyst to the possibility of a problem. Demographic factors include at least the following categories:

- Racial composition of both the targeted group and any comparison group or groups.
- Income characteristics. Ideally, the income factor should be further disaggregated by age and race.
- Education characteristics. The education factor should also be disaggregated by age and race.
- Percentage of home ownership. This measurement should be disaggregated by age, race, income, and education. Percentage of home ownership is often used as a surrogate for the level of political activism and influence of a particular population group. For example, comparing the differences (after controlling for income and education) in home ownership across two or more populations or community groups can provide insights on resistance to some siting proposals.
- Length of current residency. Length of residency is important in determining the sequence of some site location events. Did the sites follow the demographic pattern, or did the reverse occur? That is, without some longitudinal information, one cannot ascertain whether populations were drawn to a site by the low housing prices associated with preexisting environmental hazards nearby or whether the people were there first and companies with environmentally burdensome facilities followed them.
- Area density. It is often useful to know and incorporate the difference in population density between rural and urban areas. It is possible that for some situations, the frequency of site location may be most correlated with level of population density, or at a minimum, deeply influenced by such density. As a precaution, any analysis should at least take this factor into account.

There are, of course, other demographic features that can be measured, but the above groupings provide a useful first step. The guiding principle in choosing which demographic measurements to use, if any, is always to select those that make operational sense. For example, measuring differences across racial groups must also include controls for income, perhaps education, and even area population density. Ignoring these other influence factors significantly reduces the importance of any race-based differences.

Exposure Factors

Although exposure factors may not apply in many environmental-justice situations, for many others, they definitely apply. One of the first exposure factors to consider is to distinguish between the mere presence of conditions that represent potential health risks and a potential hazard. At the same time, it is important to distinguish the type of environmental media in which a threat exists. Large amounts of a certain toxin in media such as air, water, or soil may represent a lower actual risk than small amounts of the same toxin in another medium.[5] In general, the most significant exposure factors are as follows:

- Type of hazard.
- Level of each hazard.
- Medium of containment.
- Exposure routes (how populations came in contact with the hazard).
- Level of risk.

Although difficult to measure, the actual risk to the community's health should be assessed, especially when the danger is high. In such cases, epidemiological information should be collected.

Realistically, information about these exposure factors may in many cases simply be impossible or prohibitively expensive to collect. Nevertheless, even an awareness of the importance of exposure factors in evaluating some environmental-justice situations has value.

Policy Factors

Although not actually a measurement issue, questions of what type of regulatory or policy issues are involved in a particular environmental-justice problem can affect the approach an analyst adopts. Although all potential environmental-justice problems deserve attention, higher priority should be given to some. In the order of priority, the most immediate action is required if the environmental-justice problem is completely or partially caused by one or more of the parties violating existing major environmental laws or regulations. Slightly less pressing are problems that result from the violation of minor or secondary statutes or regulations. In both these situations, the operational task is made much easier because a clear legal foundation for action is present.

Potentially as important as a major environmental regulation violation, but much more difficult to get a handle on, are practices or conditions that do not violate any environmental statutes or regulations but may nevertheless violate other major federal statutes, such as the Civil Rights Act or the Fair Housing Act. In such a case, the EPA may be responsible only for reporting the violation to another agency. Realistically, in many such cases agency responsibility cannot be directly assigned.

For yet other policy situations, the EPA does not have primary responsibility for a problem, and environmental statutes or regulations have not been violated. Such situations include determining the acceptable level of toxin in consumable fish (a primary responsibility of the Department of Agriculture) or ascertaining the quality of total suspended particles in the runoff from a surface-mining operation (for which the Department of the Interior would be chiefly responsible). Here the issue is not prosecution, but rather arriving at a policy that better takes differentiated population impacts into account.

A trickier situation exists when no particular agency or governing body has primacy or when responsibility is ill-defined. One large class of environmental-justice problems that could fall into this category are issues of regional or interstate bias in exposure to environmental risk. The EPA does not have a very strong mandate to protect against differences in environmental risk exposure among groups of states where neither race nor income plays a major role. Nor does a remedy come easily to mind for such differences.

SOLUTION FACTORS

Not all problems of environmental justice should get equal treatment, nor are solutions available in all cases. The following solution factors are listed in order of increasing difficulty.

- A clear and definite solution is available that requires little interagency cooperation. In some situations, for example, preventing the occupancy of a site, while perhaps legally difficult, is a clear solution.
- A solution is available, but it requires significant cooperation between different agencies or across different levels of government. Many problems will fall into this category.
- A solution is available, but responsibility for its implementation rests primarily with government below the federal level.

- A solution is less obvious, but it appears amenable to further study and analysis.
- No direct solution is obvious at any level of government. Even after considerable analysis, the solution, if one exists, may involve a radical restructuring of society itself, or major redefinitions of the social contract.

The above prioritization is not rigid, but it offers one method for delineating courses of action. What it provides is a convenient method for avoiding the confusion resulting from failing to recognize that differences in solution may represent significant differences in policy approaches.

Methods of Measurement

Direct measurement of even the indicators of environmental-justice problems is not always possible. In some situations, investigation must proceed without quantification. However, the following operational guidelines will be useful for those problems for which measurement is possible.

MEASURING GEOGRAPHICALLY SPECIFIC PROBLEMS
First and foremost, the appropriate level of spatial analysis must be used. In other words, some geographically specific environmental-justice problems should be analyzed at the county level, others at the city level, and yet others at the level of the census block group. There is no hard-and-fast rule. The underlying policy question should dictate the choice of level of analysis. Many of the present problems in the spatial analysis of environmental justice are due to the choice of inappropriate spatial units. For most assessments across "communities" or neighborhoods, the census tract or census block group is the most appropriate level of assessment. Only for regional or interstate comparisons may the county be the preferred level of assessment.

Actually, in many assessment situations, the appropriate approach is to use more than one level of spatial unit for the analysis. In this way, the analyst has an opportunity to test how sensitive results are to changes in such levels of comparison. That is, if an environmental-justice problem is indicated at the census tract level of comparison but not the county level, additional sensitivity analysis may be warranted.

For community assessment of geographically specific environmental-justice problems, distances from exposure risks are measured, not occurrences within a fixed region. That is, the number of potentially hazardous sites—such as Toxic Release Inventory or Superfund sites—should not be counted within a city, a census tract, or even a block group. Rather, the number of sites within a specific distance from an identifiable community point or points should be measured. For fair assessment, several distances should be employed (e.g., sites 0.5, 1.0, 2.0, and 4.0 miles from the center of a census tract or block group). It is dangerous to rely on a single-radius assessment. Be especially wary when environmental-justice conclusions change radically with minor variations in the distance chosen. This is a clear indication that more careful and varied distance measurements need to be taken. Even these precautions do not take into consideration topographic variations—hills, valleys, even building patterns—all of which may affect the distribution of exposure risk.

People face risks, not land. Thus, for policy purposes, it may make sense to define a neighborhood or community as a collection of (whole and partial) census tracts. The "community" thus defined may not correspond to a preexisting boundary definition. However, this "community" may correspond to the viable level of policy-making and as such should be incorporated in any analysis. That is, populations within the boundaries defined by this collection of tracts or block groups may share a similarity of risks.

MEASURING DEMOGRAPHIC FACTORS

Both race and income are often major demographic factors in environmental-justice situations. The challenge, however, comes in devising a useful method for distinguishing between categories of these two criteria. A method sometimes used is to assign all neighborhoods or communities above or below a given racial percentage or income level to a particular category. For example, all communities with a percentage of African Americans greater than the national (or state) average are identified as minority communities; all communities whose average income is below a certain level are identified as low-income. Although there are occasions when such an approach may have merit, relying exclusively on a simple fixed percentage as a means of assessing minority or low-income status is poor analysis and provides even worse policy guidance. Such a simplistic approach obscures all other information by creating an essentially binary, either–or measurement from what may be more appropriately evaluated as points along a continuum.

In the area-specific assessment of any community, whether a small neighborhood or a much larger multicounty region, a flexible approach should be taken. Each particular situation should be evaluated within its individual parameters. However, the following basic guidelines will be useful.

When deciding if a community merits minority or low-income status, it is often best to compare that community's demographic percentages with those of the next-largest areal unit. For example, income level or racial composition percentages for a census tract or block group should be compared with the percentages for the city or county, not against some national standard. Thus, an African American population of 10 percent in a community in Phoenix would indicate a very different condition from the same percentage for a community in Washington, D.C. Because the average percentage of African Americans in Phoenix is considerably below 10 percent, a reading of 10 percent for a given community there would probably justify labeling it a minority community. In Washington, D.C., a population figure of 10 percent African Americans for a community would indicate exactly the opposite.

The base and comparison units of percentage analysis should be routinely varied when evaluating the status of a community. For example, does the status change if zip codes are used instead of census tracts? Is the status different depending on whether the community is compared with another area in the same city or to other urban areas across the country? An affirmative answer to either question is a clear warning that more detailed assessment is necessary before drawing any conclusions about environmental-justice conditions.

As I admonished earlier, wherever possible, classifying communities into distinct categories should be avoided. Instead, perform any analysis along a continuum. For example, when assessing the degree of minority exposure to potential risk, use the population percentage of the minority or the level of income as a variable in a regression analysis, rather than a dummy variable or binary designation (0 or 1) of minority status. If a binary approach must be adopted, be sensitive to the inevitable accompanying loss of information.

Tools of Analysis

Given the vast range of environmental-justice problems, no single methodology can satisfy all analytic demands. The availability of information and the underlying analytic objective will largely dictate the specific choice of method. For many questions of environmental justice, the paucity of data can severely limit that choice. Yet even when a wealth of situational information and other data can be obtained, the wisest course of action may not always be to turn to the most powerful—and usually most difficult to execute—analytic tool. For many initial investigations, a simple comparison of percentages, with no statistical testing at all, may suffice. On the other hand, when a crucial issue is at stake, if at all possible different methodologies should be applied to the same problem and the results and conclusions compared.[6]

Methodologically, environmental-justice problems can be analyzed not only by the more traditional evaluation techniques of the social sciences, but also by such wide-ranging disciplines as geographical information systems, the large area of risk assessment, and finally some parts of operations research and management science. Thus, many analysts, trained in only one branch of methodological investigation, may not be up to the task. A thorough investigation of an environmental-justice problem may require teams of analysts, each approaching the problem from a different perspective.

To complicate things further, many environmental-justice issues do not lend themselves to any form of quantitative analysis. Many of these issues, such as a Native American decision to accept increased environmental risk by having tribal lands serve as a nuclear waste depository, require a nonquantitative substantive assessment and are intimately connected with philosophical issues of self-determination and alternative economic outcomes. For this wide array of nonquantitative environmental-justice questions, the facts of a situation are not in question, only the underlying conclusions.

Although the evaluation of each environmental-justice problem requires its own methodology, certain general principles apply in nearly all cases:

- Fit the methodology to what the data will allow.
- Fit the methodology to the level of analysis or type of decision-making required.
- At all times, recognize the limitations of the chosen methodology.
- Avoid choosing a methodology on the basis of its difficulty or complexity.
- Realize that just because a methodology is popular does not make it right for the problem at hand.

Table 8.1 suggests broad, admittedly arbitrary categories into which the methods of problem analysis may be divided.[7]

Qualitative Decision-Making and Analysis

These methods include a wide range of techniques, from Delphi methods for consensus discovery to gaming exercises. The manipulation and analysis of data assume a secondary role to the judgmental and intuitive assessment of a problem. Such methods are most useful for making a normative or moral assessment when numbers are secondary. They are also useful when a paucity of information prevents the application of more quantitative methods. Even when data are available, however, these methods can prove useful for scenario-setting and problem structuring. Actually, the first step in the analysis and measurement process, regardless of what data are available, is identifying the major parameter of the problem. The qualitative approach can often serve a crucial function in helping this identification and structuring process.

Table 8.1. Methods of environmental justice analysis.

TYPES OF PROBLEMS	TYPES OF METHODS					
	Qualitative analysis	Descriptive statistics or numerics, including GIS	Univariate and multivariate statistical analysis, risk assessment	Multiple regression-related econometric methods	Computer simulations and other operations management procedures	Operations research techniques such as DEA for assessing multiple-risk conditions
Evaluation of a normative outcome or making an ethical assessment	**Best choice;** includes Delphi Method	Useful for background information only	Not useful	Not useful	Useful only as a scenario for background information	Not useful
Descriptive assessment of community or population conditions	Can provide useful secondary information	**Best choice;** percentages, averages, and similar descriptive statistics	**Best choice;** risk assessment of community and population conditions	**Good choice;** identifying contributing factors for community conditions	Not necessary	Not necessary
Static comparative analysis of two or more communities or populations for a single risk or environmental condition	Use when few or no data are available; results are not definitive	Useful only if better data or methodology are not available	**Best choice**	**Best choice**	If data are available, not necessary	Not necessary
Dynamic comparative analysis and risk assessment of two or more communities for a single risk or environmental condition	Use when data are missing or more quantitative methodology is not available	Of some use as discrete point in time descriptors; would include time change graphs, etc.	**Best choice;** if data are available, a powerful tool for some dynamic processes	**Good choice;** time-series regressions useful for analyzing some dynamic processes	**Good choice;** when data are not available or process is too complicated, simulations are a good alternative	**Possible choice;** may not be necessary
Static comparative analysis and risk assessment of two or more communities for multiple risks or environmental conditions	Some use for descriptive background purposes	Not very useful	**Possible choice;** some forms of multivariate analysis can accommodate multirisk comparisons	Not very useful	**Possible choice;** if data are not available for other alternatives	**Best choice;** power operations research tools are useful for multirisk condition comparisons, assuming data are available
Dynamic comparative analysis of two or more communities for multiple risks or environmental conditions	Some use for scenario building	Some use for descriptive highlights of points along the dynamic line	Not very useful	Not very useful	**Possible choice;** may be only way to assess dynamic dimensions of a multirisk problem	**Possible choice;** under some conditions can provide insights into parts of dynamic conditions

Descriptive Statistics or Numerics

These procedures will often precede more powerful quantitative engines or serve as a supplement to qualitative assessments, but they can also stand alone. Basic statistics—such as averages, percentage of population engaged in an activity, or level of income—provide information necessary for the assessment of a condition. Virtually none of the more complicated procedures can proceed without them. On the whole, this approach is not only the best choice for a descriptive assessment of communities or populations, but it also provides the basic data ingredients for more complicated static and dynamic comparative procedures.

For both single and multiple dynamic risk comparisons, descriptive statistics performed at several points in time is a simple way to highlight dynamic changes, although it falls far short of actually providing a dynamic analysis. Geographical information systems—with their associated linkage between areal location and economic, financial, demographic, and other information—is the key methodological step in many environmental-justice analyses. Epidemiological studies, while usually not quantitatively complicated, are often the preferred way to identify risk vectors and to separate out the various factors contributing to a community's environmental risk.

Univariate and Multivariate Statistical Analyses

These broad classes of quantitative evaluation, more than any other method reviewed here, provide the backbone of environmental-justice analysis. They are the best choice for most descriptive assessments of conditions within a community or population, including the application of risk assessment. The decision analysis forms of statistical analysis are the best choice for evaluating a sequential or multistage decision process.

These methods are also one of the best choices for performing a static comparative analysis of two or more populations, states, or conditions. The powerful tools of hypothesis testing and statistical inference permit the examination of competing explanations of the forces that drive existing conditions. In the form of multivariate techniques such as canonical correlation, these tools even have some utility in evaluating and comparing multiple-risk situations. They are least useful for assisting in dynamic, multiple-risk evaluation. Of course, no other method can provide much assistance with such dynamic problems either.

Multiple Regression and Related Economic Methods

Multiple regression, which actually is a subcategory of multivariate statistical analysis, offers in its many forms powerful estimation capability for identifying factors that contribute to a condition. Multiple regression is an excellent choice for a comparative static analysis of a single risk factor or single dependent factor across populations with varying demographics or other characteristics. For example, this is the most likely choice when one is comparing incidence of location of a single type of hazardous-waste facility across a collection of communities to differences in racial composition of the community. Time-series versions of regression analysis can help in assessing dynamic processes. However, without an extraordinary amount and range of information, including interactive effects, regression analysis does not give much assistance in examining multiple-risk conditions. This assessment applies to both the static and, especially, the dynamic forms of multiple-risk evaluation.

Computer Simulations and Other Operations Management Procedures

This category of methodologies has not enjoyed much use in the general environmental policy area. Because these methods originate from a discipline unfamiliar to many environmental policy or program managers, they are little understood or appreciated. For some of the more complicated environmental-justice programs, these methods may offer the only opportunity for evaluation. A maxim of management science is that if you do not have enough information about the process to do anything else, or if the process is too complicated for a real analysis, use a simulation.

For more normative outcomes or decision-making, these methods can provide some assistance in constructing the background scenario. And for extremely complicated processes, such as multiple risk assessment, especially over time, a computer simulation may be the only method for the assessment of outcomes. Other operations management procedures, such as queuing theory and forecasting, can also be useful for some environmental-justice evaluations.

Operations Research Techniques

The final methodology category, although definitely not the least powerful, is that of operations research. Techniques such as linear and goal programming do not at first appear applicable to any environmental-justice problems. However, as I will explain in more detail in Chapter 9, some operations research approaches may be the best, if not the only way, to perform multiple-risk comparisons. On the other hand, operations research is usually not very useful for normative or descriptive assessment, even for background information purposes.

One technique within operations research, data envelopment analysis, can be used to perform some multiple-risk comparisons of population groups. Linear programming can serve as a maximizing or minimizing tool to discover the range of some constrained alternatives. Although not perfect, some forms of operations research, such as a time-based version of data envelopment analysis, can provide results that shed light on parts of a multiple-risk, dynamic comparison.

Conclusions

To a large extent, the problem should dictate the method, not the other way around. Applying an inappropriate methodology is worse than doing nothing at all. Such applications give the illusion of objective rigor where none exists and offer misleading conclusions and directions. Too often in policy analysis, one encounters a "have method, will travel" mentality. The greatest strength of good policy analysis does not necessarily come from the computation, evaluation, or predictive power of an available set of methodologies. Rather, its greatest strength may often be the organizing and structuring of the problem that must precede the methodological application. In many environmental-justice problems, no methodology can provide the answer directly. But the very act of walking through a seemingly intractable problem and identifying its key elements and dynamics will reap benefits. Particularly for environmental-justice analysis at its current stage of development, structuring has just as much importance as methodological application.

Nine

A New Way of Looking at the Same Old Numbers
*Using Data Envelopment Analysis to
Evaluate Environmental Quality*

*The fact that most studies focus on one risk factor at a time some-
times supersedes all other measurement problems in the field of environmental justice. Unfortunately, in the real world, risk factors often
come in groups. It follows, then, that evaluating the status of environmental justice in a community by assessing a single risk factor is
equivalent to evaluating the performance of a university solely by
the number of publications written by the faculty in a year.*

Statistics are seldom exciting, except to a statistician, and even then only on
alternate Mondays. Many readers may feel exempted from wading through
this chapter after having forded the previous chapter on information and
data analysis. That chapter explored the many data and information pitfalls
faced by any analyst, community group, or even curious investigator ready
to tackle the environmental-justice question. Why, then, do we need a
chapter that probes these problems further and involves even more mathe-
matics? Besides providing the equations that are obligatory for all books on
policy analysis, this chapter introduces and explores a new and powerful
method called data envelopment analysis (DEA), which offers a unique
way of evaluating a number of environmental-justice questions. The reader
who has no desire to acquire such knowledge is encouraged to skip all but
the first and last few pages of this chapter. But please read at least those
pages to gain some sense of the vital insights this approach can provide.

It may seem pointless to explore yet another analytic methodology. In
many areas of environmental-justice analysis, numbers are clearly subordi-

nate to issues of behavior or politics.[1] In fact, it may be argued that the spirit of the environmental-justice movement is the rising environmental awareness and sense of injustice felt, expressed, and acted on by disenfranchised groups. Nevertheless, while recognizing the importance of spirit, the fuel and building blocks of this movement are the information and revelations gleaned from careful analysis of environmental conditions across and among various populations and communities. Environmental policy today is very much a game of numbers and equations. Bad numbers, unreliable statistics, and inadequate methodologies lead to bad analysis, unreliable conclusions, and inadequate policy solutions. Issues of environmental justice must be viewed from a firm analytic and methodological foundation. Failure to adopt such an analytic perspective guarantees that the environmental-justice movement will become nothing more than another campaign of empty slogans.

The last chapter explored some of the controversies surrounding the collection and interpretation of data and information in this drama. But even if all problems involving data and information were resolved, the problem of weak methodology would still remain.

Current methods of analyzing the environmental-justice issue of unfair community environmental risk or burden are inherently misleading. Until now, most analytic investigation of environmental equity problems have employed simple multivariate statistics or similar quantitative comparisons of variations in a single-risk measurement against a collection of explanatory community demographic measurements or characteristics. This is the equivalent of evaluating a university strictly on the basis of the publication numbers of its faculty. Although clearly important, and definitely a factor in overall university quality, faculty publication counts provide no information on many other important dimensions of university performance. In the same way, measuring and comparing occurrences of a single risk factor such as community location near Toxic Release Inventory (TRI) sites or near hazardous-waste management sites, while clearly important, definitely do not accurately portray the entire environmental risk picture a community faces.

In the more ambitious of the single-risk studies, environmental inequity has been identified as occurring when race continued to explain differences in levels of potential risk, even after researchers controlled for income, education, and other characteristics.[2] Although not invalid, the conclusions of such studies are seriously weakened because their associated methodologies are unable to consider multiple sources of risk simultaneously.

Why is the methodological limitation that allows evaluation of one risk factor at a time such a big problem? What is lost by this limitation? Suppose a category of communities is identified by simple cross-sectional comparison[3] as having a disproportionate number of hazardous-material risk sources when compared with other communities within some larger area. Furthermore, assume that this study arrived at this conclusion after trying to explain variations in site frequency by variations in community income, race, education, and similar demographic characteristics. Finally, assume that race was the most powerful explanatory variable for this phenomenon. Surely then, in such a case, we must have an example of environmental justice. Are not these communities being subjected to disproportionate environmental risks?

The answer may be yes. But the answer may also be, Who knows? For unless there is a high correlation between a particular risk factor with all other environmental risk measurements relevant for this community, such a broad conclusion as race-based environmental inequity cannot be drawn solely from the results of an analysis that is based on a single risk factor. It is possible, for example, that although site location may indicate an unequal risk burden for this one particular factor, other risk factors, such as the quality of air and groundwater, may paint a very different picture. By the same token, when studying another measurement, such as air quality, as a potential risk factor, if we do not encounter an uneven pattern of risk distribution over the same set of communities, we do not have conclusive support for the conclusion that there is *no* race-based environmental inequity.

With a few exceptions, most early risk analysis concerned itself with evaluating the risk to individuals created when a particular risk factor entered or was altered in an environment. Such factors included smoking (or being exposed to the smoke of) a certain number of cigarettes a day, or exposure to a given level of PCB (polychlorinated biphenyl) contamination over a given period of time. Such analyses could also be reframed in terms of the distribution of risk across population groups. For example, which income or racial groups are most affected by cigarette smoking?

Unfortunately, many environmental-justice issues have to be considered in a context that does not lend itself to assessment via the single-risk-factor approach. For these factors, the "what" is less often, "What is the distribution of risk factors across population groups of risk factor x?", but rather, "What is the total or aggregate difference in environmental risks of many types that a particular population or community faces?" Knowing that population A faces a higher exposure to risk factor x does not truly answer the question, "What is the total environmental make-up of risk factors x_1, x_2, through x_n, that population A faces?"

This chapter explores an alternative approach for assessing environmental conditions in particular communities or geographical areas. This alternative is not a replacement for other statistical procedures that are properly employed to assess some categories of environmental-justice issues. This new procedure simply provides an additional tool in the analytic arsenal, one useful for assessing multiple-risk situations. Given the current sorry state of both methodologies for environmental risk assessment and the availability of appropriate data, no single method of analysis should ever be solely relied on to provide policy insight.

Furthermore, locational inequity is just one (although an important one) of a range of environmental-justice issues. For many of these other issues, such as pesticide exposure among Hispanic farm workers, neither current bivariate or multivariate statistical procedures, nor the alternative proposed in this chapter, have much relevance. In many cases, an environmental-justice issue is not one of discovering whether a pattern or distribution exists. Rather, such issues often demand a value judgment about whether a particular circumstance is an environmental inequity that requires remedial action or whether it is simply an unavoidable outcome of market dynamics.

A major problem with many previous statistical approaches to environmental equity in which only one source of environmental hazard at a time is assessed is that the differences between racial groups for one risk factor may be offset by the distribution of other risk factors among the groups.[4] In contrast, the method presented here is borrowed not from classical statistics but rather from the field of operations research. This method applies DEA, the powerful estimation mechanics of a technique from the frontier of management science, for an evaluation of the relative potential exposure of various populations to an amalgam of environmental hazards.

By using DEA in this completely new way, to segregate the evaluated community groups along racial lines, racial differences in potential risk from multiple sources of pollution can be assessed simultaneously. An investigator can thus obtain a fairer, clearer assessment of community status and determine the likelihood of bias operating in the process.

Previous Studies in Environmental Quality/Justice

In earlier works done in the area of environmental quality or justice assessment on the specific issue of TRI sites or hazardous-waste facility location, the general conclusions have been that nonwhite communities endure a disproportionate share of potential environmental risk that cannot be explained away by differences in education or income.[5] The authors in all cases employed some form of multivariate statistical analysis to compare the occurrence of incidences of a single environmental risk phenomenon across geographical units. In all these various studies, this occurrence of a hazard or risk factor is compared against a collection of community demographics such as race and income for each geographical unit.

Two categories of serious measurement and methodological questions, although not questioning the validity of such works, have been raised about these works. The first category comprises concerns regarding the choice of geographic measurement unit.[6] The second group of concerns relates to the choice of a single-factor modeling approach to answer issues involving multiple risk factors.

As to the first group of questions, consider that several of these works, such as the pioneering study by the United Church of Christ,[7] used postal zip codes as their basic geographic unit, whereas others, such as the work by John Hird, used the even larger unit of a county for community base measurement.[8] And this can be a problem. As noted by several others—for example, D. Anderton et al.[9]—even in urban areas, postal zip code zones are fairly large and, worse, do not correspond to the boundaries of census information such as census blocks or tracts.

Furthermore, the use of the county as the basic unit of comparison is even more curious because in an urban area, the assumption of uniform population distribution across an entire county is especially hard to support or accept. In an area such as Los Angeles, the county approach assumes that race and income are essentially the same from Beverly Hills to Watts. An interesting study by W. Bowen et al.[10] that used data from the Cleveland area showed that by expanding the unit of measurement from the census tract to the county, the conclusions of inequity can actually be reversed—from inequitable for nonwhite communities to inequitable for white middle-class communities.

Typical of the community-level work in examining environmental quality or risk patterns is a study by Vicki Been[11] that reexamined work on environmental and race conditions in Houston, finding for the same point in time results similar to those of a much earlier pioneering work by Robert Bullard.[12] In both cases, a disproportionate share of hazardous-waste sites were located in nonwhite communities, even after controlling for income. The Been study, however, took the next step and focused on whether the facilities' location decisions preceded or followed the nonwhite residential concentration pattern in the Houston area. That is, were the neighborhoods first nonwhite and then had such facilities move in, or were the majority of the facilities there first and then nonwhite citizens simply took advantage of low land values and moved into the community? This is more than an interesting social science observation. If facilities were attracted to a community because of its minority status, as opposed to minorities being attracted to a community due to the low cost of housing (which in part may have been due to the proximity of environmentally hazardous facilities), two very different policy positions emerge. In the first case, a strong argument for environmental racial bias exists; in the second case, we observe a market dynamic that does not necessarily require a policy response other than acceptance. The study results suggest that the former rather than the latter occurred in the Houston case, giving support to a bias in locational decision-making rather than a simple market phenomenon for this community.

As to the second group of concerns, in several of these earlier works, the single-focus context of the basic environmental-justice question makes policy sense. For example, Bullard's and Been's studies on Houston focused exclusively on whether hazardous-waste facilities were disproportionately located in minority communities. In a similar manner, the Anderton et al. study explored the same issue across a national framework. In these three studies, the key issue was not whether minority communities were disproportionately affected by environmental hazards or policies in general. The question in these studies was whether decision-making about the location of these particular sites could be racially biased. The answer for Bullard and Been was probably yes, and for Anderton, no. Not pausing to debate the merits of these individual conclusions, the methodologies employed in those single-risk studies were valid. When in such studies as these, or at least part of such studies, the focus is carefully limited to issues in which asking questions about a single risk factor has some policy significance, concern over the multiple-risk issue loses some of its importance.

Where we encounter problems is when the results of such works as those cited above are used as the foundation for broad conclusions about overall environmental inequity in a community or among a population. Ignoring issues of proper measurement units—whether to use counties, zip codes, or census tract or block groups—it becomes too much a leap of faith to conclude on the basis of a single factor's characteristics that environmental-justice problems do or do not exist in a community or across a region. Yet this is precisely the leap of faith that many have attempted.

The DEA Approach

DEA was developed in the late 1970s by Abe Charnes, William Cooper, and me as a way to measure how well a particular program or operation performs relative to other programs performing similar tasks.[13] Such a relative evaluation is obtained via a DEA estimation of a single or scalar measurement of performance efficiency.[14] This value ranges from 0 (impossibly inefficient) to 1.0 (the unit under assessment is located on the relative efficiency frontier). Values in between suggest how far off the frontier an investigated unit was. Thus, a DEA efficiency value of 0.75 loosely indicates that the unit in question is only about 75 percent as efficient as those units that make up the relative best performance frontier.

DEA uses a variant of linear programming to evaluate the effects of a set of input, or influencing, factors on a set of output, or outcome, factors. In fact, in a manner quite parallel to the environmental-justice problem of evaluating single- versus multiple-risk factors, DEA was originally developed as a means of assessing multiple-output processes that other methods could not easily handle. This new evaluation yields an individual scalar measure of performance efficiency that is calculated by the DEA algorithm for each unit or operating system being evaluated. The resulting relative efficiency measures, in turn, can be used in a second-stage analysis via multivariate statistics. This later-stage statistical analysis of the initial DEA results allows examination of assorted influences that could have induced the variations in DEA values across observational units.

As a management science alternative to more traditional econometric methods of productivity analysis, DEA has been particularly successful in public-sector applications. This is not surprising because in this arena, multiple-output processes are common, if not the norm. DEA has been applied in public-sector operations as diverse as education (see Charnes, Cooper, and Rhodes;[15] Charnes, Cooper, and Ahn;[16] Bessent and Bessent;[17] Rhodes and Southwick[18]), health care (Sherman;[19] Banker, Charnes, and Cooper[20]), criminal justice (Lewin, Morey, and Cook[21]), natural resource management (Rhodes[22]), public utilities (Fare, Grosskopf, and Logan[23]), public finance (Adolpson, Cornia, and Walters[24]), and military operations (Bowlin[25]). In all these cases, the public-sector activities involved multiple outputs. And in all these cases, DEA was chosen as the assessment approach because there were no viable alternatives.

DEA's growing success can be traced to several analytic and computational features. These features are especially noteworthy from the perspective of policy analysts and decision-makers participating in the environmental-justice dialogue. First, from an analytic perspective, DEA is one of only a few estimation procedures that can assess multiple outcomes. So for multiple-outcome analysis, the methodological choice must be DEA or one of a small set of alternative, and usually more computationally difficult, statistical procedures.[26] What is equally attractive—and what places DEA ahead of most of these other methods—is that a DEA estimation makes the fewest specifications for data and modeling of any of the multiple-outcome evaluation procedures.[27]

The second attraction of the DEA approach is that DEA is computationally straightforward. That is, it does not take a Ph.D. in operations research or computer science to set up and run a DEA calculation. Of course, this does not address the challenge of choosing the proper size of measurement unit that is faced in environmental-justice assessment, but to a large extent, that is less a methodological problem and more one of proper matching between evaluation objectives and the attributes of a particular geographical measurement unit. That is, for some investigations, the census tract may be the best choice, yet for other investigations, a larger unit—the county or even a region—may be more appropriate.

There now exist several relatively inexpensive DEA software packages for both mainframes and microcomputers that make entering and running a DEA calculation simple. The literature on DEA has grown sufficiently large that explanations of most common computational problems can be readily answered.[28]

But what we explore here is not the relative performance or efficiency of a set of public or private service producers. This study takes DEA in a new direction. Here, DEA is employed to assesses the relationship between a collection of influences of the conditions in a community—such as average level of education or average income, or wealth, race, and similar demographic characteristics—and an associated collection of potential environmental risks—such as number of hazardous-waste sites in the community, number of TRI sites, number of Superfund sites, or average air quality.

We begin from the premise that race—or whatever target bias one chooses—does not play a major role in determining the location outcomes of a particular set of environmental risks. Then, after taking into account, via DEA, the influence of other demographic factors, such as income, education, and the percentage of home ownership in the community, we would expect our DEA-estimated values for community quality to be randomly distributed across geographic areas, with no correlation with race. If, however, race appears as a significant explanatory variable in explaining the variations in DEA values across this set of community groups, then an argument of racial environmental inequity gains credence. In the same manner, one could look at income as a bias factor. In this case, all the demographic and economic factors would enter the DEA calculations, and it would then be seen whether the excluded factor of income variation could explain the resultant variation in DEA-estimated values for these communities.

This use of DEA differs from all its many past applications. In most previous research, the emphasis has been on identifying a relative activity or production efficiency frontier. Typically, this frontier identifies the most efficient from among a group of producers or providers of a good or service. When evaluating environmental conditions, however, there is no interest in or attempt at identifying a frontier of activities. Nor can the phenomenon being evaluated even be described as a production activity. The idea here is to identify those communities that, although they enjoy favorable demographic characteristics (income, education, etc.), still face a relatively high potential for environmental risk. The interest in such an evaluation is thus in impact, not economic efficiency. In this context, it is possible that randomness alone will account for some outcomes, but if it is observed that nonwhite communities are collectively "inputting" relatively high demographic factors but are realizing high environmental risk, environmental inequity may exist.

As noted earlier, computationally, the DEA mechanic is a twist on the standard linear programming maximization problem. A nonlinear ratio of a collection of environmental quality factors[29] is evaluated over a collection of demographic-influencing factors for each of a collection of individual geographic areas such as census tracts or census block groups. The objective is to maximize the value of the ratio of the weighted vector of environmental quality factors over the weighted vector of demographic factors. The unknown coefficients are the values of these vector weights. This maximization problem faces the constraint, as in the original DEA formulation, that no individual set of ratios of weighted outputs to weighted inputs can have a value greater than 1.0. That is, no community can exceed a relative best environmental quality of 1.0, or 100 percent. This is a relative evaluation because units are only compared with other units within their cohort. This means that at least one unit must have a DEA score of 1.0 in any standard DEA calculation.[30] Units with values below 1.0 are viewed as in some way not performing at their relative best. Thus, in the case of environmental risk exposure, a neighborhood with a DEA value below 1.0 is not enjoying the level of environmental quality we would expect from a community with its demographic characteristics. In such a neighborhood, some unmeasured factor must be preventing or retarding better environmental quality.[31]

The resultant calculated DEA values would then be used in a second-stage statistical analysis as a dependent vector with race and other previously nonincluded demographic or social characteristics used as explanatory variables. In the second-stage statistical analysis, if race proves significant as an explanatory variable for the deviation in DEA values, then a conclusion of environmental inequity may be justified. That is, inequity may be present if, for example, the minority percentage in a community, or similar measurement of community minority status, helps explain why that particular community has a DEA value for relative environmental quality of less than 1.0. As we should remember, a DEA value of less than 1.0 does not imply that a community is doing poorly in an absolute sense, but rather that, given that community's demographic characteristics, a higher environmental quality index was expected.[32]

DEA Environmental Illustration

The attraction and method of DEA may become clearer if we walk through a small illustration of a hypothetical community. The following illustration

Table 9.1. Characteristics of sample communities.

Census tract	Average income ($)	Average education (grade level)	Average value of home ($)	Population density (per square mile)
CT$_1$	5,000	9	25,000	300
CT$_2$	10,000	13	70,000	450
CT$_3$	8,000	10	60,000	1,000
CT$_4$	10,000	11	85,000	800
CT$_5$	5,000	8	65,000	200
CT$_6$	6,000	8	60,000	100
CT$_7$	8,000	10	45,000	900
CT$_8$	5,000	11	50,000	200
CT$_9$	7,000	8	30,000	400
CT$_{10}$	8,000	12	30,000	500
CT$_{11}$	7,000	10	25,000	300
CT$_{12}$	10,000	12	65,000	100
CT$_{13}$	7,000	9	50,000	800
CT$_{14}$	9,000	11	70,000	600
CT$_{15}$	8,000	10	70,000	800

will not only demonstrate how relative environmental status is calculated by means of DEA, but will also show the dangers of drawing conclusions that are based on an analysis of a single risk factor. For this example, assume that there are fifteen census tracts in a given community: CT$_1$ through CT$_{15}$. For analyzing these tracts, let us select five community characteristics as input (or influence) factors, of which four will be included in the initial DEA calculation. The fifth, minority percentage, will be added during the second-stage statistical analysis. Two measurements of environmental quality have been chosen as outcome factors. The four initial demographic characteristics are census tract–specific: mean household income, median years of school completed, median value of owner-occupied homes, and mean population density per square mile within the census tract. Table 9.1 shows the values for each census tract.

The above observations would then be entered as the DEA input or demographic influence factors. To get the environmental quality measurements, we begin with two environmental risk measurements: number of TRI reporting sites within a mile of the geographic centroid of the census tract, and a measure of air pollution within the census tract.

Unlike the demographic characteristics, the above numbers will not be directly entered into the DEA calculation. DEA operates directly only for positive outcomes or outputs, and thus, the risk factors that are negative measurements have to be converted. We measure each environmental quality factor variable as the difference plus one from the largest risk value of its type within any census tract.[33] For each risk factor measurement, this becomes

$$y_{rj} = (\text{Max factor}_r - \text{factor}_{rj}) + 1$$

where y_{rj} is the environmental quality measurement of the rth factor for the jth neighborhood; Max factor$_r$ is the maximum value for the rth risk factor scored by any neighborhood in the sample; and factor$_{rj}$ is the value of the rth factor for the jth neighborhood.

Thus, for a given collection of hazardous-material sites, the larger the number actually entered into the DEA formula, the better (as opposed to worse) that risk condition is for that neighborhood. Consider the measurement of TRI hazardous-material sites as recorded in Table 9.2. The largest possible number of sites any neighborhood could have within its measured radius of 1 mile is 5.0. Thus, the conversion CT_{11} would have a site outcome value of $(5 - 2) + 1 = 4$; CT_7 would have a site value of $(5 - 1) + 1 = 5$; and CT_{15} would have a site value of $(5 - 4) + 1 = 2$. The same principle would apply to the second environmental measurement, air pollution. The largest air pollutant value is 45. For air-quality measurement, then,

Table 9.2. Exposure risk of sample communities.

Census tract	No. of hazardous sites within one mile	Air pollution index
CT_1	1	23
CT_2	5	12
CT_3	2	45
CT_4	2	23
CT_5	4	11
CT_6	3	5
CT_7	1	30
CT_8	1	18
CT_9	3	23
CT_{10}	4	12
CT_{11}	2	15
CT_{12}	0	20
CT_{13}	3	26
CT_{14}	0	24
CT_{15}	4	6

Table 9.3. Environmental quality of sample communities.

Census tract	Hazardous sites within x miles	Air pollution index
CT_1	5	23
CT_2	1	34
CT_3	4	1
CT_4	4	23
CT_5	2	35
CT_6	3	41
CT_7	5	16
CT_8	5	28
CT_9	3	23
CT_{10}	2	34
CT_{11}	4	31
CT_{12}	6	26
CT_{13}	3	20
CT_{14}	6	22
CT_{15}	2	40

CT_{11} would have an air value of $(45 - 15) + 1 = 31$; CT_7 would have an air value of $(45 - 30) + 1 = 16$; and CT_{15} would have an air value of $(45 - 6) + 1 = 40$. These converted numbers would therefore be as in Table 9.3.

If we first had run a DEA that, as with the more common statistical method, only analyzed one risk factor at a time and chose only the number of hazardous-material sites as the risk measurement, the resultant estimated DEA values of relative environmental quality would be as in Table 9.4.

Now that we have fifteen DEA values that provide some measurement of relative environmental quality, what do we do next? We are now ready to introduce the race factor via a second-stage statistical exercise. Assume now that for these fifteen census tracts, the percentage of the nonwhite population in each neighborhood is as Table 9.4 lists.

With the above information on minority percentage in a census unit, we can proceed in several ways. For example, we could perform a simple linear regression with the computed DEA values for relative environmental quality as the dependent variable, and the minority percentages as the independent variable:

$$DEA_j = f(Min_j)$$

Table 9.4. DEA results for sample communities for a variety of variables.

Variable	DEA results for single risk factor (no. of TRI sites)	Proportion of minorities in sample communities	DEA results for sample communities for two risk factors
CT_1	1.00	0.20	1.00
CT_2	0.14	0.64	0.61
CT_3	0.72	0.60	0.72
CT_4	0.65	0.12	0.69
CT_5	0.47	0.50	1.00
CT_6	0.75	0.45	1.00
CT_7	0.90	0.40	0.90
CT_8	1.00	0.20	1.00
CT_9	0.67	0.60	0.84
CT_{10}	0.33	0.40	0.94
CT_{11}	0.83	0.10	1.00
CT_{12}	1.00	0.20	1.00
CT_{13}	0.69	0.30	0.67
CT_{14}	0.98	0.50	0.98
CT_{15}	0.36	0.70	0.81

where DEA_j is the calculated DEA value for relative environmental quality for the jth neighborhood and Min_j is the minority percentage in the jth neighborhood.

Looking at the results of the regression, if the racial percentage variable explains a significant amount of the deviation in the DEA values, we could conclude that race appears to be a factor in explaining levels of environmental quality beyond that already explained by other demographic characteristics such as income and education. If we ran such a linear regression for the illustrated values, we would get the following results:

Sample regression results with one risk factor:
$$DEA_j = 1.0021 - 78.07 \text{ minority}$$
$$(7.267) \quad (-2.479)^*$$

(t statistic; *statistically significant at the 0.05 level).

The illustrated regression result clearly suggests, with a t statistic of -2.479, that a negative relationship exists between level of environmental quality and racial percentages across census tracts.[34] Apparently for this illustrative group, as minority percentage increases, the level of environmental quality decreases, even after controlling for the influence of income, education, and wealth (housing value is a common surrogate measure of wealth). Now, obviously other influencing factors—such as differences in underlying housing preferences between racial or ethnic groups or compensating side payments for particular neighborhoods to endure increased environmental burdens—are not captured and could also account for this difference. Nevertheless, solely on the basis of these simple results, one could conclude that a problem of environmental justice may be present here and require government intervention or assistance.

Another method of approaching the equity issue would be to divide the observed tracts into two categories, those with a minority percentage above the area average, and those below the average, and then compare their DEA values to determine whether there is a significant difference between the two. For this simple illustration, the average minority value is 0.394. Thus, tracts CT_2, CT_3, CT_5, CT_6, CT_7, CT_9, CT_{10}, CT_{14}, and CT_{15} have minority populations above the average, and tracts CT_1, CT_4, CT_8, CT_{11}, CT_{12}, and CT_{13} have below-average minority percentages. The mean DEA value of the first set is $DEA_{above} = 0.6237$, and of the second, $DEA_{below} = 0.7857$. We observe, then, about a 26 percent higher DEA mean for the census tracts with below-average minority percentages. This result would further support the existence of environmental inequity.

Is this the correct conclusion? For the illustrated hypothetical values, is there truly strong evidence of environmental inequity? The results of the single-risk-factor approach suggest yes. But is this sufficient? Consider what happens when a two-risk-factor DEA calculation is performed. For this run, the converted values for both number of TRI sites and air quality are included as environmental quality variables. The resultant DEA values appear in Table 9.4.

Performing yet another simple regression, we get the following somewhat different results:

Sample regression results with two risk factors:
$$DEA_j = 0.9707 - 0.2844 \text{ minority}$$
$$(11.638) \quad (-1.254)^*$$

(t statistic; *not statistically significant at the 0.05 level)

In contrast to the earlier single-risk-factor results, these results suggest that although a negative relationship has been detected between DEA value change and minority percentage in a census tract, the results are not statistically significant and may be due to nothing more than random chance. The results in this case argue against a major problem of environment equity.

Performing the same secondary simple comparison of census tracts with above- and below-average minority percentages, the results are that average value for $DEA_{above} = 0.857$ and average value for $DEA_{below} = 0.900$. The DEA_{below} group mean is just 5 percent larger than the DEA_{above} average, not a significant difference. Thus, an examination of both the simple regression results and the group means strongly suggests that there is no serious discrepancy in environmental conditions within this sample community due to racial differences.

Summary Observations

The extra effort of employing the DEA approach in this illustration has paid off. With DEA, which allows a more comprehensive consideration of the risk factors, the picture of environmental burdens in this community changes considerably. Without DEA, such a consideration would not have been possible. In the illustrative case, in fact, very different conclusions emerge when the analysis is based on two risk factors rather than only one. Given the political and sensitive nature of this topic, mistakes of analysis and improperly formed conclusions are definitely to be avoided. DEA, while not a cure-all, unquestionably provides an avenue for clearer understanding of a class of environmental-justice situations.

Realistically, though, will all outcomes change, as was the case with this illustration, when more than one risk factor is considered simultaneously? Probably not! In many cases, there will probably exist some degree of correlation across risk factors. In those cases, a significant finding with one risk factor probably does carry some generalizable significance. However, the correlation is not perfect, and ignoring the multidimensional nature of the factors facing a neighborhood diminishes one's ability to argue for or against the existence of an environmental-justice problem in a community.

Although the above illustration described a community at the census tract level, the DEA approach has just as much applicability at much broader levels of analysis. For example, if the question is one of regional bias in the distribution of certain risks, the DEA mechanism can just as easily accommodate units of analysis such as counties or states in making broad comparisons across multistate regions. This is also a good moment to point out again that DEA does not directly resolve the question of the proper size of the community selected for analysis. In many cases, the optimal strategy is to investigate the issue at several levels of aggregation or disaggregation. If the results radically change as one moves from, say, using census block groups to using zip codes or similar larger geographic units, the question at hand is not choosing one measurement size or another. Rather, the more important question is what factors may have contributed to this outcome shift and whether those factors have any policy significance. In most cases, the size issue is not one of proper choice of evaluation methodology but instead one of choosing the proper community unit size to fit the policy questions being explored. And in some cases, the proper answer may be to provide results from several different size analyses and simply point out the contributing factors (if you can identify them) that lead to changes in results. Conversely, in the case of results that are stable across a range of measurement unit sizes, note the improved confidence one can now assign to the results.

Returning to the DEA methodology, in all these applications, whether at the smaller community level, the much larger regional level, or something in between, the most important fact to remember is that DEA by itself does not prove or disprove anything. DEA provides an insight into the behavior of some condition. That insight is still dependent on the quality of the initial information placed into the DEA algorithm—garbage in, garbage out. DEA, which can process multiple-outcome situations, does not in any way reduce the need for carefully constructed measurements of those outcomes. The use of, for example, too wide a radius of inclusion when making a site count has the potential for distorting the significance of a risk factor or environmental condition, and DEA will in no way solve or correct this problem. Furthermore, the DEA results must still undergo second-stage statistical analysis to be of any real assistance. This latter point cannot be overemphasized. DEA results work best when integrated into a larger analysis exercise in which there is ample opportunity to contrast or supplement DEA results with other information.

Another consideration totally ignored in DEA and any multivariate statistical analysis is that even if such distributional distortions are present and in fact are admittedly based on race or income, there may be compensating factors. What neither approach can directly capture is the possibility that the location or risk concentration was a deliberate and voluntary decision on the part of the community for some form of compensation. It is possible, and for current and future location decisions likely, that communities have received a counterbalancing collection of payments to offset the increased potential risk they face. Even in these cases, however, it would be wise to engage in a DEA study to provide a benchmark of comparison. Find out how different a community's conditions are and decide whether the compensation is sufficient.

Last, a reminder: the DEA approach only applies to the class of environmental-justice problem that involves questions of distributional distortions over geographic areas or social categories. By and large, the DEA approach works best in situations in which the alternative would be some form of multivariate statistical analysis. This naturally means that for large classes of environmental-justice questions—such as the dietary patterns of some minority groups that expose them to a higher level of toxins, or the higher incidences of pesticide exposure among farm workers—neither multivariate statistics nor DEA are appropriate.

Measuring Environmental Inequity

Given the recent genesis of interest in environmental justice, neither the quantity nor quality of current data on environmental hazards is well suited to careful investigation of the subject. In fact, risk assessment in general has not progressed to the point where the measurement of the cumulative effects or magnitude of risks of a single impact, and especially of multimedia impacts, can be made for an individual or a community. For this reason, in this study—as in most studies in the area of environmental risk assessment—when I speak of risk exposure, I mean potential risk exposure rather than actual risk exposure, because the latter is unknown and to date unmeasurable. Nor have I attempted to aggregate any collection of risk factors to obtain a single-risk summary.

Even if one accepts the current limitations on risk assessment, other serious data problems remain. For example, the data available on environmental exposure are not well integrated with community demographic information. Because environmental policy-makers had until recently ignored race- and income-related impacts or policy issues, their management information systems and data collection practices were not formulated to easily provide such information. Furthermore, serious questions remain about the reliability of much of the location-specific environmental data (one estimate suggests a minimum error rate of 25 percent). In other words, good community-specific assessment of environmental equity depends on accurate information regarding environmental activity within a given space. Unfortunately, numerous errors can be found in both the specific location of hazardous sites and the exact area covered by a site. Furthermore, errors continue in the self-reported levels of pollutants released by specific sources.

This study cannot hope to address even some of these issues of analysis and evaluation. My ambition for this chapter was limited to demonstrating that for one dimension of this complex issue, the use of an assessment tool such as DEA allows useful insights. At the same time, I hoped to raise sensitivity to the overall problem of measurement that this class of environmental-justice issue raises. I intend to introduce a way to assess differences in the aggregate average risk that different communities may face due to the presence of various environmental hazards. Of particular concern here is identifying patterns of potential risk distribution that are not random but rather follow some identifiable configuration associated with a particular set of community characteristics such as race or income. Of course, we must not forget that identifying such differences does not itself prove causality. Much more is needed than what is provided via a DEA application, however refined, to establish a causal link to decision-making bias. Nevertheless, the results of a DEA analysis that indicate a racial or income bias definitely bolster the argument for potential environmental inequity far more than do results from a community assessment employing a single-risk approach.

MATHEMATICAL APPENDIX

A note for those filled with questions about how one could get a single scalar measure of efficiency from a set of two vectors of multiple outputs and inputs. Briefly, the scalar efficiency measure is a ratio of the weighted output vector over the weighted input vector. The real question then is, How do you determine the weights? And the real answer is, You do not. The weights are objectively determined within the DEA calculation process itself. How? Basically, the weights are those values for each output and input that will maximize the ratio of outputs to inputs, given the constraint that no ratio value for any individual sample point or unit can be greater than 1.0. In other words, no unit or community can be more than 100 percent efficient. Mathematically, this can be depicted as

$$Max\ z_o = \frac{\sum_{r=1}^{w} \mu_r y_{ro}}{\sum_{i=1}^{m} v_i x_{io}}$$

subject to

$$1 \geq \frac{\sum_{r=1}^{w} \mu_r y_{rj}}{\sum_{i=1}^{m} v_i x_{ij}}$$

x_{ij} = amount of input factor i observed in the jth unit or observation
y_{rj} = amount of output factor r observed in the jth unit or observation
v_i = objectively determined weights on input i
μ_r = objectively determined weights on output r

and

x_{io} and y_{ro} are the specific input and output values for a unit "o" chosen from the set of n observations sampled for evaluation of a particular DEA analysis run

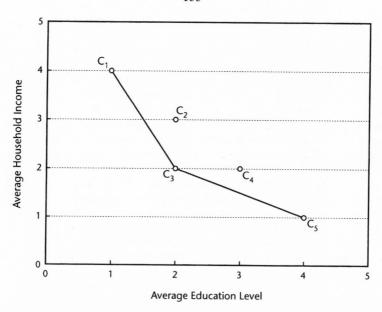

Figure 9.1. Relationship between income and education in five communities.

For those who have been carefully following the above discussion about objectively determined weights, it should be obvious that what results is a Pareto definition of efficiency. That is, the calculated weights obtained are those that give each unit the best efficiency value it could possibly achieve given its combination of output production and input usage. This means that there could exist some "theoretical" true efficiency value for this unit that would be considerably lower in value (i.e., closer to 0). Note, however, that no theoretical value could ever be greater than the calculated DEA value. We thus say that the DEA value is an extremely forgiving approach for calculating efficiency. So although we cannot be certain that every unit on the relative best performance frontier deserves to be there, there is no question that any unit with a DEA value less than 1.0 definitely deserves to be off the frontier.

Graphically, this set of relationships can be depicted for the extremely simple case of the evaluation of five communities with one environmental quality factor and two demographic influence factors—say, average level of education among adults and average income. We further simplify things by specifying that for the set of five communities being investigated, the value of the environmental quality outcome is in all cases 1.0. Specifying a common value for the environmental outcome permits us to then depict both influence factors in a two-dimensional graph. Figure 9.1 represents a sample set of five communities. In this depiction, education level is on the horizontal x-axis and average household income on the vertical y-axis. In this figure, the object is to get as close to the origin as possible—that is, to have achieved the environmental quality level of 1.0 with the greatest relative efficiency (i.e., the lowest) for the combination of education and income levels.

In the arrangement depicted above, C_1, C_3, and C_5 are relatively efficient, with DEA values of 1.0. That is, they make up the relative environmental quality frontier. In contrast, C_2 and C_4 have DEA values less than 1 (actually, in both cases, the DEA value is 0.857). Note that C_2 is viewed as inefficient because it has as high a level of education as C_3 and a higher level of average income than C_3 yet enjoys the same level of environmental quality. In the same way, C_4 has as high a level of income as C_3 and a higher level of education, yet again only enjoys the environmental quality outcome level of 1.0. Conversely, C_1 actually has the highest average income, but this does not prevent its location on the relative frontier because it also has the lowest level of education.

Of course, the above figure depicts an extremely simple example. When we move to the true multiple-outcome/multiple-input case, things become graphically much more complex. So although relationships with multiple influences and multiple environmental conditions cannot be depicted in a two-dimensional graph, the relationship can be formally depicted algebraically in the same manner as the original DEA equation. That is, the DEA for calculating relative environmental conditions becomes a maximization of a ratio of weighted outputs over weighted inputs, constrained by the conditions that no individual community can have a ratio greater than 1.0. As before, this is specified as

$$Max \; z_o = \frac{\sum_{r=1}^{w} \mu_r y_{ro}}{\sum_{i=1}^{m} v_i x_{io}}$$

subject to

$$1 \geq \frac{\sum_{r=1}^{w} \mu_r y_{rj}}{\sum_{i=1}^{m} v_i x_{ij}}$$

and $\mu_r > 0$; $v_i > 0$
for $r = 1...,w$; $r = 1...,m$; and $j = 1...,n$

x_{ij} = amount of influence factor i observed in the jth community
y_{rj} = amount of environmental quality factor r observed in the jth community

as before

v_i = objectively determined weights on input i
μ_r = objectively determined weights on output r

x_{io} and y_{ro} are the specific influence and environmental quality values
for a community
"o" chosen the set of n communities sampled for evaluation
in a particular DEA analysis

However, in this case, the vector of x's are inputs of various community characteristics such as mean household income and level of educational attainment, and the y's make up a vector of resultant environmental quality factors based on a conversion of a set of environmental risk measurements. That is, the environmental quality measurements will be conversions into positive terms of a collection of negative-oriented environmental risk factors. For example, the original environmental risk factors could be the number of TRI sites within a given geographic range and the number of National Priority List (NPL) sites from the larger Superfund site list within a certain area of a neighborhood.

Next, choose the maximum number in each category of TRI sites and NPL sites observed for any community area in the observation set. Now, from the maximum number, subtract the actual number of TRI and NPL sites observed in each community. In this way, a community with no TRI and no NPL sites would have an environmental quality measurement for two factors that were maximally high. In contrast, those communities with sites would have their environmental quality number reduced from the average.

There may be, for the astute reader, some concern over the apparent assumption that an environmental quality measurement of, say, 8.0 for TRI sites is twice as good as an environmental quality measure of 4.0 for NPL sites. But this is actually not true. One of the other attractions of the DEA approach is how it handles factors of varying scales and magnitudes. What DEA requires at a minimum is an assumption that 8.0 represents within its own scale of TRI a better level, although not twice as high, as a value of 4.0 for another on the TRI-converted scale. It makes no assumption about the relationship between specific values of TRI and NPL sites. If, however, through a priori information one has additional relationship information that actually does spell out the relative importance of TRIs versus NPLs (such information at present being highly unlikely), the DEA model does have the capacity for modifications to reflect these additional bits of information.

Part III

A Case, a Summary, and Some Conclusions

To appreciate that there are rarely any "obvious" cases, Chapter 10 explores in some detail one particular environmental-justice case involving the siting of a hazardous materials facility in the community of Noxubee County, Mississippi. Noxubee County is significant only in that it is typical of a broad category of environmental-justice issues. In the Noxubee case, there were three sides: the landowners, the hazardous-waste facilities company and its community allies, and community opponents of the facility and their allies. Note that two of the three sides claimed to speak for the community. All three players had different objectives and perspectives on the issue. Each side had some valid points, and each side had some less than proud moments during the controversy.

Returning to the question of government involvement raised in Chapter 11, even accepting that such action is justified, there still remains the issue of the nature and scope of that involvement. Examples of questions that could be raised about the form of involvement are: Should the principal form of government involvement be direct legislative mandates or laws prohibiting certain actions, or should it adopt regulation criteria based more on performance criteria? In such regulation, compliance is not the completion or execution of a specific set of actions, but rather the satisfaction of a specific set of outcome standards by whatever action one chooses. But if based on performance criteria, how are such criteria arrived at? Rejecting the performance criteria approach, what about a series of regulations outlining specific methods of guaranteeing environmental justice in the conduct of specific environmental activities? Such an approach would mean a massive regulatory undertaking to codify every environmental activity with environmental-justice implications. Given the extremely inefficient and largely mixed results of the Occupational Safety and Health Administration's regulation of the American workplace, will a similar approach to environmental justice result in a similar mountain of regulations and very little justice?[1] Nevertheless, circumstances may dictate that such a command-and-control approach to government intervention is the only viable option. Which way to go?

Ten

A Case of
Environmental Justice

*The Disposal of Hazardous Material
in Noxubee County, Mississippi*

As a problem in environmental equity, the case of Noxubee County, Mississippi, is noteworthy for its ordinariness. An underinformed minority population appears to be the potential victim of both environmental and economic exploitation by a giant industrial enterprise trying to ensure its own economic well-being. Also playing a role in this drama are institutions as varied as a long-established community civil rights organization opposing its state chapter, a major midwestern university that made a questionable land decision, a state government anxious to boost regional economic development, a federal government agency anxious to avoid a lose–lose engagement, local commerce groups opposing economic development, mainstream environmental organizations trying to gain credibility in the environmental-justice area, and finally, student organizations from the same midwestern university newly discovering environmental racism. What develops is that in such a complex scenario, where all outcomes have both a positive and a negative side, labels of "good guy" and "bad guy" prove difficult to assign. The "right position" is a vague place that everyone professes to occupy, but to which no one has an exclusive claim.

The Major Players

Noxubee County

Located along the northeastern edge of Mississippi, sharing a state line with Alabama, rural Noxubee County has no major industry and high long-term unemployment. The county, a nearly perfect square, is traversed by only one major highway, U.S. 45, which runs north and south through its middle. Approximately 45 miles north of Meridian, Mississippi, and about 15 miles north of Emelle, Alabama,[1] Noxubee County, 695 square miles in area, has a population of about 12,600, of which 2,256 (18 percent) live in the county seat of Macon. The rest of the county is sparsely populated; it is the eleventh least dense of Mississippi's eighty-one counties. For several decades, Noxubee, like many small rural Southern counties, has experienced a net loss in population.[2] The largest other towns in the county, connected to Macon by U.S. 45, are Brooksville (population 1,098), near the north-central border of the county, and Shuqualak, near the south-central border (with fewer than 900 residents). Over two-thirds of Noxubee County's population (68.6 percent) is African American (in 1990, Mississippi was 35.6 percent African American).[3]

Among all the states, Mississippi ranks near the bottom in educational attainment; fewer than two-thirds (64.3 percent) of people older than twenty-five have a high school diploma. Noxubee County fares even worse; fewer than half (49.6 percent) of its citizens older than twenty-five are high school graduates, and only 7.9 percent have earned a bachelor's degree. Adult illiteracy in Noxubee County is estimated at approximately 45 percent. Among the 68.6 percent of the county's population that is African American, for those older than twenty-five, only 14.3 percent are high school graduates.

In Noxubee County, 41.4 percent of households live at or below the U.S. poverty level, and the county has one of the highest unemployment rates in the state.[4] In 1990, the county had a median household income of less than $14,205, well below the Mississippi median of $20,136. But although just 6.7 percent of white households in Noxubee County live below the poverty level, 55.9 percent of African American households do. The median income for white households in the county is $23,198 (15 percent above the median for Mississippi), whereas African American households have a median income of $10,411 (48.3 percent below the state median). Stated another way, Noxubee African Americans have a per capita income of $3,869, against $12,278 for Noxubee whites. The economy of Noxubee County depends primarily on agriculture. These activities include corn and cotton farming, catfish aquaculture, some cattle ranching, and timber processing for the construction industry. Nearly all enterprises in the county, including its largest employer, a paper mill, are owned and operated by whites.

Noxubee County has a board of elected supervisors, who have primary responsibility for county governance, and several of its small towns, such as Brooksville, have elected town boards of aldermen. In Brooksville, two of the board's five members are African American. The county is also home to a local chapter of the NAACP, with about a hundred members.

The Indiana University Foundation

Chartered in 1936 as a not-for-profit corporation, the Indiana University Foundation "is designated by the trustees of Indiana University as the official fund-raising agency for the University."[5] However, as is common with many fund-raising and endowment management systems at public universities, the foundation, although assisted by the state, is not a state agency. The foundation has a close and overlapping relationship with the administration of Indiana University (IU), although it has its own governing board of directors. For example, the president of IU also serves as the paid chair of the forty-two-member board of directors of the foundation. Furthermore, several seats on the foundation's board are reserved for members of the university's board of trustees. The foundation's board is dominated by Indiana residents and former alumni of IU. The board meets as a whole three or more times a year, with committees or smaller groups holding meetings as needed throughout the year. Daily operations of the foundation are carried out by a full-time staff, directed by the foundation president, who also has a seat on the board.

As of June 1992, when the Noxubee County case unfolded, the foundation had total assets of $308,020,129. Its total income in 1992 was $48,689,362, of which $15,497,111 came from investment income and net realized gain. In 1992, its total administrative expenditures were $9,002,283, and about 25 percent of its operating expenditures ($2,300,000) came from state-appropriated funds, via a fee from IU assessed as payment for the development and fund-raising performed by the foundation for the university.[6]

Conrad Farm in Noxubee County

In the early 1970s, after having their offer of land declined by Purdue University, Martin and Opal Conrad of Indianapolis, Indiana, in a series of exchanges, gave two major properties to the Indiana University Foundation: a 1,300-acre grain and cattle farm in the Indianapolis area, and a farm and ranch of nearly 6,000 acres in Noxubee County, Mississippi. It is not clear why the Indiana University Foundation accepted the properties because the ranch never appeared to have been economically viable. In fact, soon after Martin Conrad completed purchasing the land that made up the Conrad ranch, in the 1960s, he and his wife tried to offer it to Purdue University, which declined to accept it because of the poor condition of the land.[7] Regardless of the reasoning, the foundation did accept the land with the understanding that it would eventually be sold.

Upon the death of Martin Conrad in 1988, the Indiana University Foundation began, with the approval of his widow, to look for potential buyers for the Mississippi property. When Opal Conrad died in July 1990, no obstacle stood in the way of its sale. By the time the property was placed on the market, the president of IU at that time, Thomas Ehrlich, who was also the chair of the Indiana University Foundation, had held his position for less than two years. He had not participated in and was unaware of the specifics of the original gift arrangement. Similarly, the president of the Indiana University Foundation, Curtis Simic, also an outside appointment and having only just begun his tenure in 1988, had no input into the original property arrangement. From late 1988 to September 1990, attempts to sell the property met with no success. As the property remained in the hands of the Indiana University Foundation, its financial losses continued.

After the property had been on the market for almost two years, in September of 1990, Edward Netherland, chair of Federated Technologies, Inc. (FTI), emerged as the first serious prospective buyer.[8] Netherland was forthright in explaining that he proposed to use the land as the site of a hazardous-material landfill and incinerator. On September 20, 1990, with little notice to or comment from the university community, the foundation approved granting FTI an option to purchase the property on or before September 1991 for $3.5 million.[9] This amount was $1.5 million more than the Indiana University Foundation's estimate of the land's value as a ranch or farm.

FTI and Their Proposal

When it proposed to build a hazardous-waste disposal facility in Noxubee County, FTI was a new corporation with no previous experience constructing or managing a hazardous-material disposal site. The CEO and chair, Edward Netherland, had come from the insurance industry. Several of FTI's senior managers, however, had a number of years of experience in the waste-management industry. Originally chartered early in 1989 in Tennessee, the company's financial statement as of December 31, 1989, showed total assets of $10,105, and its facilities consisted of a rented office in Nashville.[10]

At its founding, FTI had as its president former Tennessee Health and Environmental Commissioner James Word, and Edward Netherland, then president of Netherland and Associates, was chairman of its board. The new company sought two Tennessee permits and a site for constructing and operating a hazardous-waste incinerator in that state. The Tennessee project was to employ a hundred workers and occupy between 40 and 100 acres of land. According to the original proposal, the site would not include a hazardous-waste landfill. FTI actually bought land in Giles County, Tennessee, and sought unsuccessfully to acquire a 107-acre plot in Macon County. Plans at the two proposed sites in Tennessee met with opposition from both local and state officials—including then Governor Ned McWherter—and were either abandoned or put on hold.

In Mississippi, FTI incorporated as Federated Technologies of Mississippi, Inc. There has been some disagreement over exactly when the Indiana University Foundation was originally contacted by FTI about the land purchase, as well as disagreement about who in Noxubee County knew of their interest in and intended use of the land before the foundation's official announcement. FTI stated that they had held more than thirty meetings with interested citizens of Noxubee County, but only from June 1990 on. Nevertheless, by September 1990, FTI had reached a formal agreement with the foundation to purchase the 6,000-acre Conrad farm, contingent on FTI obtaining the necessary state permits.

Sensitive to local questions about the project, FTI—under the signatures of chair Edward Netherland and president Fred Wynn, a chemical engineer—issued an open letter to the citizens of Noxubee County. In the letter, FTI outlined the features of their proposal:

- This would be a "high-tech waste management facility."
- Sixty million dollars would be spent building the facility.
- When it became operational, the facility would provide more than 180 jobs.
- The facility would have an annual payroll of $4.5 million.
- An estimated $10 million would be spent on local purchases once the plant was running.
- The facility would be "100 times more effective than the most stringent EPA [U.S. Environmental Protection Agency] standards."
- The hazardous-waste incinerator would use the rotary-kiln method, which produces a much safer final product than alternative methods.
- The Tennessee-based environmental engineering firm of DRE Technologies, Inc., would assist in the site's design and construction.[11]

During the discussion that accompanied the siting process, FTI stated that only three factors had influenced their choice of the Noxubee County location:

- The site was one of four identified by the state as suitable for a hazardous-waste facility. The estimated 650 feet of Selma chalk under the Noxubee County site made it "probably the most impermeable and deep formation in the country."[12]
- The site had access to railroads and modern highways.
- The site was big enough to provide a large buffer zone.

By mid-October, FTI had notified the Mississippi Department of Environmental Quality (DEQ) of their intentions as Part A of their application process: to file a complete application for a site permit by February 1991.[13] The notification included a proposal to use about 100 acres for the actual facility and another 1,300 acres for a buffer zone. The remaining 4,600 acres would be used for agriculture or for residential and industrial development.[14]

Other Players

Not long after FTI's initial proposal, two other companies expressed an interest in locating waste-processing facilities in or near Noxubee County. On January 17, 1991, U.S. Pollution Control, Inc. (USPCI), of Houston, a division of the Union Pacific Railroad, notified the Mississippi DEQ of their plans to apply for a permit to operate a hazardous materials operation in southern Noxubee County, off Hwy. 45 near Shuqualak, about 12 miles south of the proposed FTI site. This proposed facility would have three landfills, two liquid treatment plants, and possibly an incinerator. The liquid treatment plants would include wastewater treatment and a solvent recovery operation. The three separate landfills would be for hazardous waste, industrial nonhazardous waste, and municipal waste.

In November 1990 yet another firm, Recycling Sciences International, held public meetings in Lowndes County, Mississippi—Noxubee's northern neighbor—outlining plans for a dirt-recycling plant in Artesia, less than 15 miles from FTI's proposed site in Noxubee County.

FTI Moves Ahead

In pursuing its plans for a hazardous-waste disposal site, FTI obtained agreements or endorsements from several governmental and nongovernmental bodies in the county. The most significant of these was reached on December 11, 1990, with the Noxubee County Board of Supervisors. Subsequent agreements were signed with the Town of Brooksville, the City of Macon, and the Town of Shuqualak on February 5, 1991. The agreement with Noxubee County included, among other provisions, the following:

- FTI agreed to pay the Board of Supervisors a fee of $1 million after the granting of a construction and operation permit by the state.
- The parties agreed that 30 percent of all monies received by the State Tax Commission from FTI's Noxubee operation (as tonnage handling and disposal fees) would go to the general fund of Noxubee County, of which 30 percent of the FTI tax money to the county would go to the three municipalities involved in the project: Macon, Brooksville, and Shuqualak.
- FTI agreed not to resell permits to a third party for at least five years from the date they were issued.
- FTI agreed to provide the county with a solid-waste landfill for a period of twenty years on the same or nearby land. This landfill, however, although operated as a not-for-profit venture, would still require collection of fees, 50 percent of which would go to FTI.[15]
- After the permit or permits were granted, FTI would provide matching funds, up to $125,000, for the building of a county civic center.
- Noxubee County and the towns of Macon, Brooksville, and Shuqualak would agree not to allow any other hazardous-material site to be located in the county. (This stipulation was modified when USPCI then proceeded with their permit plans.)
- These same governmental entities would take steps to be designated collectively as a "volunteer host community" under the Mississippi Hazardous Waste Siting Act of 1990.

On February 5, 1991, FTI made additional agreements, of which the four most relevant were as follows:

- The City of Macon's solid waste fee to either FTI or Noxubee County would be limited to $8,500 per year.
- In addition to state fees, FTI would pay $1.00 per ton of hazardous material handled to the three municipalities, which would divide this revenue equally among themselves.
- To compensate for increased infrastructure demands, the three municipalities would each receive $50,000 per year from FTI.
- FTI, as part of what they would call the Good Neighbor Program, would fund a program of scholarships and charitable activities by placing $0.50 per ton of hazardous material handled into an escrow account,

up to a maximum of $100,000 per year. The principal from this account would remain part of FTI's assets, and the resultant interest would be used for the Good Neighbor Program.

FTI's Partnership with Hughes Aircraft Corporation

On March 23, 1991, FTI announced that Hughes Environmental Systems, a division of Hughes Aircraft Corporation (a subsidiary of General Motors), would be joining them in their plans to build and operate a hazardous-waste disposal site on the Noxubee property. FTI revealed that Hughes had approached them the previous year about such a partnership and that plans had advanced enough to make the announcement. By this time, the proposal also included plans for an environmental research facility that would possibly involve Mississippi State University and several other state-supported higher education units. This research facility would be called the Center for Environmental Optimization and would focus on problems and "solutions in a variety of environmental areas, including hazardous waste management facilities that employ state of the art technology in all aspects of design, development, implementation, and operation."[16]

Hughes Environmental Systems helped FTI submit to the state DEQ the crucial Part B of their application for a permit to construct and operate a hazardous-waste incineration facility in Noxubee County. The partnership was not made official until August 16, 1991, when Hughes announced the proposed facility, which, it said, "will create more than 200 new jobs and bring in an estimated $10 million annually to the area."[17] It also appears that the overall facility, rather than just the research component, would bear the name Center for Environmental Optimization. Hughes also noted that its partnership with FTI should "calm fears that FTI could not run such a technical operation."[18]

The facility would have the capacity to incinerate up to 50,000 tons of waste per year. Its three landfills would also be able to accept up to 300,000 tons of hazardous waste per year. FTI's own statements made it clear that as a minimum, the company was basing these capacities on the needs of all of EPA Region IV rather than on the needs of Mississippi alone.[19] In fact, given the current operation of several large-capacity hazardous-waste disposal sites in this region, the area served may have extended even farther.

A brochure, under the Hughes name, released by the partnership between Hughes Environmental Systems, Inc., and Federated Technologies of Mississippi, Inc. (HESI-FTI), identified three major components to the Center for Environmental Optimization: a Research and Development

Facility under the "guidance of the Mississippi university system," an engineering facility focusing on technology development and design called the Advanced Technology Facility, and a Comprehensive Industrial Waste Treatment Facility, which would include both the hazardous-waste incinerator and landfill. The brochure stated that the capital investment would be more than $100 million for construction and start-up. It also reiterated the earlier claim of $10 million in annual local spending or revenue infusion. HESI-FTI further announced the proposed establishment of a "vocational training program to provide local residents with necessary skills to fill 80 percent of the more than 225 jobs" that would be created by the project. Annual payroll was estimated to be between $5 and 6 million, with an average wage of $12 per hour and a minimum of $7.

Requirements of the Resource Conservation and Recovery Act and the Superfund Amendments and Reauthorization Action

The process of obtaining a permit for siting a hazardous-waste incinerator or landfill came under the federal 1976 Resource Conservation and Recovery Act (RCRA). As amended in 1984 and 1986, subtitle C of the RCRA, a cradle-to-grave system of accountability was specified covering original-source generators, hazardous-waste transporters, and the final disposal site. The act stipulated safety provisions for a new or expanded old site, such as special double plastic liners for containment, leachate collection systems, and monitoring of surrounding groundwater. The amended act also established the permitting and enforcement powers of the EPA over the location and operation of a site, including the power to invoke both civil and criminal penalties.

The RCRA also provided that states may assume responsibility for the overseeing of the handling of hazardous materials, with the caveat that any state laws less stringent than RCRA requirements would be superseded or preempted. Actually, the EPA was never prepared for, nor was money budgeted for, monitoring the nation's hazardous wastes on a site-specific basis; from the beginning, it encouraged the states to assume this responsibility. Mississippi was given the responsibility for monitoring its own hazardous wastes in 1984.

Where a state has assumed enforcement of RCRA requirements, the EPA has very little direct responsibility for, or authority over, the selection of a hazardous materials site. Even in states where the EPA has retained direct authority for granting operating permits for a site, the exact location of a particular site and the decision to permit building still remains with local government.

Contained in the 1986 reauthorization act for the Comprehensive Environmental Response, Compensation, and Liability Act—called the Superfund Amendments and Reauthorization Action (SARA)—was a requirement that each state assure the EPA that it had sufficient capacity for managing hazardous waste generated within its borders for twenty years after submission of that assurance. This capacity assurance requirement could be satisfied either by demonstrating sufficient capacity in facilities within the state or by having access to facilities in other states through interstate or regional "agreements." Failure to do so would result in a loss of Superfund support for nonemergency cleanups.[20]

Nowhere in the SARA's requirements for capacity assurance is there a provision that a state must satisfy the capacity assurance plan by building and operating a facility *within* its borders. In 1989, Mississippi passed a set of laws, supposedly based on the SARA provisions, that specified procedures for dealing with waste management, including the disposal of hazardous waste. The governor followed this up in 1990 by appointing a group to find a site for facility construction and operation within the borders of Mississippi. Contrary to the actual requirements of SARA, the state of Mississippi has maintained that it was carrying out the mandates of SARA requiring such a facility within each state when it insisted on aggressively pursuing approval of the Noxubee site.

The State of Mississippi

In 1990 state legislation, ostensibly passed to satisfy the handling capacity requirements of the Comprehensive Environmental Response, Compensation, and Liability Act, SARA Title II, mandated selection of a site and construction of a hazardous-waste facility thereon. In September 1990, the governor, Ray Mabus, appointed a Hazardous Waste Siting Authority, which was given fifteen months to find such a site in the state. FTI's proposed site would have met this requirement without requiring the state to assume responsibility for disposal.

The state government, especially the governor's office, wanted the site for a disposal facility to be decided on and the facility placed into operation as soon as possible. The FTI plan not only fulfilled this intent, but offered the added incentive of placing most of the obligation for capital investment and operating expenses into private hands, relieving the state of an unwanted fiscal burden. FTI's plan also had the potential for stimulating development in Noxubee County's otherwise stagnant economy.

Support and Opposition in Noxubee
Community

As soon as, and possibly even before, the original option to purchase the Conrad property was announced in September 1990, opponents of the site plan emerged. Leading the opposition in Noxubee County to the facility's construction and operation was Protect the Environment of Noxubee County (PEON). PEON, whose members may at one time have numbered nearly a thousand, had been formed five years earlier to oppose the siting of a chemical landfill in southern Noxubee County, near Shuqualak. The group not only blocked that siting but also helped obtain approval of a five-year moratorium on all landfills in Mississippi.

PEON presented the following arguments against the FTI proposal:

- Noxubee County was chosen as a site because its population—poor, uneducated, and primarily African American—was the least likely to successfully oppose their plans.
- Such deliberate environmental racism would benefit a white-owned and -operated corporation from outside the county. If the population of Noxubee County had been predominately white, it was unlikely that FTI would have chosen it.
- The agricultural and dairy activities on which the county's economy mainly depended would be jeopardized by the operation of a hazardous-waste disposal facility. This is especially true of dairy products, which would encounter extreme resistance in regional markets if consumers knew they were produced within the shadow of such an operation.
- The construction of a residential or industrial area, as FTI had proposed, adjacent to a hazardous-waste incinerator was unsafe and unfeasible.
- Incineration was an unproven and potentially dangerous method of disposing of hazardous waste. Noxubee County should not be a laboratory experiment.
- The potential risks of hazardous-waste incineration were great enough that they could not be justified by the economic benefits derived from the facility's operation.
- FTI itself was an unproven company, with questionable financial capacity, which to date had neither built nor operated any type of facility. At the time of the current proposal, FTI was involved in unsuccessful and controversial efforts to build hazardous-waste facilities in two different locations in Tennessee.
- As for economic development and employment, few of the promised jobs

would go to current residents of Noxubee County, especially its African American population. Except for a few menial maintenance and labor positions, most of the FTI operation required workers with technical training and experience, which they would likely import from outside the community.

- Given the large capacity of the incinerator—50,000 tons per year and the only estimated 25,000 tons of combustible organic waste generated in Mississippi annually—FTI clearly planned on their Noxubee facility serving as a multistate regional, if not national or international, site for the disposal of hazardous waste.
- Noxubee County is only 15 miles from the nation's largest hazardous-waste facility in Emelle, Alabama, operated by Chem Waste, Incorporated. The location of a similar facility in Noxubee would only add to the region's problems. As one long-term resident put it, "We are terribly concerned about this. Our area is already called 'the pay toilet of the country,' and now we're going to be totally saturated with hazardous waste."[21]

FTI was also accused of meeting secretly with various white and African American political figures in the county as early as spring 1990 to arrange unopposed endorsement of their plans. It was charged that politicians and other community leaders were being bought off by FTI. PEON claimed that many of the procedures by which these meetings were held were improper and that the politicians who participated in them were not representing the will of the citizens of Noxubee County.

During the fall and early winter of 1990 and 1991, PEON unsuccessfully campaigned to prevent endorsement of the FTI proposal by the various governmental units in Noxubee County. On December 10, 1990, the Noxubee County Board of Supervisors voted 3 to 2 to support the FTI proposal. This vote included the county's approval of the hazardous-waste incineration and disposal plan and joint participation by the county and the towns of Brooksville, Macon, and Shuqualak in a solid-waste management program. The county's two elected African American supervisors both supported the measure. In February, after that vote, the aldermen of the three towns voted their approval of the plan by margins of 3-2, 3-2, and 4-1, respectively. The other four elected African American officials in the county—two aldermen in both Brooksville and Macon—all supported the proposal. The local chapter of the NAACP, with about a hundred members, along with several other community groups, also officially endorsed the FTI proposal.

Although PEON was not successful in preventing endorsement of the proposal by the county and town governments, they had, however, gained the attention of several environmental action organizations in the region. Greenpeace and the Environmental Congress of Arkansas both sent representatives to Noxubee County. On February 4, 1991, two days before the endorsement votes by the aldermen of the three Noxubee towns, the representatives addressed Noxubee citizens opposed to the plan.

Aware that local governmental endorsement would be inevitable, some of those opposed to the sale turned their attention toward other strategic avenues. On that same day, February 4, a petition with reportedly 2,249 signatures was sent to the Mississippi DEQ protesting the sale. Opponents also tried to dissuade the Indiana University Foundation from selling to FTI. They hoped that the state might deny the permit to FTI, or that the Indiana University Foundation might be persuaded to back out of the sale altogether.

Debate at IU

Soon after the original option to buy was established in September 1990, PEON contacted members of the Indiana Chapter of the Student Environmental Action Coalition (SEAC). Also that fall, before the endorsement of the siting by the county and the three towns, some Noxubee County citizens expressed their opposition of the sale to the foundation. However, the first significant (and publicized) contact between the community and the university did not occur until spring 1991. In April 1991, with the sponsorship of the Bloomington local SEAC chapter, Martha Blackwell, a leader of PEON, came to IU's main campus in Bloomington to speak to a crowd of several hundred students during an Earth Day celebration. For most students at the gathering, this was the first time they had heard of the Noxubee problem.[22] Sandwiched between the performances of two local bands, in a campus gathering area called Dunn Meadow, Blackwell presented PEON's objections to the sale and called Noxubee County the target of "industrial profiteering." She also noted that the county's preexisting racial tensions were exacerbating the problem and that FTI was taking advantage of the powerlessness of Noxubee's people. At the conclusion of Blackwell's speech, members of SEAC promised to sponsor a petition to be taken to the Foundation.

But despite Blackwell's visit, the Mississippi land issue aroused little interest or action on the Indiana campus. When by September 1991 FTI had failed to obtain the necessary Mississippi permits for construction and operation of the facility, the Indiana University Foundation had little problem—nor had they faced any protest—in extending the option for another $100,000 to September 1992 and allowing an extension of even that deadline for an additional $50,000, if the need arose.

Not until the spring of 1992 was any visible and vocal opposition generated on the IU campus to the sale of the Conrad farm to the corporation that was by then called Hughes Environmental Systems, Inc.–Federated Technologies Mississippi, Inc. (HESI-FTI). This opposition was led by the preexisting organization, SEAC, a national environmental organization with an almost exclusively white student membership both nationally and on the IU campus, as well as by the more loosely structured Bloomington campus Noxubee County Committee, most of whose members were African American. By the spring of 1992, the IU student newspaper, the *Indiana Daily Student,* and the local city paper, the *Herald-Times,* were chronicling the situation in Noxubee County every week, and sometimes every day. Most of the student editorials attacked the university for practicing environmental racism by allowing, if not encouraging, the sale to move forward.

That spring also saw the beginning of a letter-writing campaign to the foundation by SEAC, the student Noxubee County Committee, PEON and other members of the Noxubee community, and others dissatisfied with the sale. Beyond reiterating the position of the county residents opposed to the sale, the letters and editorials in the *Daily Student* hammered away at several additional points:

- As an institution of higher education, IU—and by association the Indiana University Foundation—has a higher moral and ethical responsibility than to serve the interests of other landowners. And as such, the Indiana University Foundation would be violating the principles their funding was supposed to support by using monies obtained under less than moral circumstances to advance the condition of IU at the expense of the citizens, especially the African American citizens, of a small, sparsely populated, poor county in rural Mississippi.
- The university administration should be held responsible for the Foundation's actions and should step in to halt the sale.
- More environmentally safe uses of the land should be found; perhaps the Indiana University Foundation should donate it as a state park or wilderness area.
- In the long run, this sale could hurt the fundraising capabilities of the foundation by leading some donors to look disapprovingly at IU and the Indiana University Foundation for agreeing to such a sale.

On May 26, 1992, SEAC introduced a resolution to the Indiana University Student Association (IUSA) opposing the sale of the Noxubee property. In the resolution, SEAC reiterated most of the charges previously raised by PEON, which included the belief that Noxubee County had been chosen as a "sacrifice zone" because its population was predominately poor and African American.[23] Passed unanimously on June 4, the resolution called the decision to sell the land an example of both "racial and economic bias" and promised the active opposition of the IUSA to the sale. The IUSA and SEAC formally presented their resolution to the Indiana University Foundation at a June 13 foundation board meeting.

By now, the Noxubee sale had become one of the major issues for the university's student government. Unfortunately, the resolution and the student publicity campaign that surrounded it began at the end of the school year and reached its peak during the summer, when student interest in political causes reaches a low ebb on the Bloomington campus. Interestingly, the African American student and faculty associations on campus, while aware of the issue, never became formally involved.

During August 1992, several student groups, including representatives from the Noxubee County Committee and SEAC, traveled to Noxubee County to assess the situation. By the time they returned, their opposition to the sale had increased. Interviews with Noxubee residents reinforced their earlier suspicions about environmental racism. Their resistance further increased when they learned that the Indiana University Foundation had obtained the mineral rights to the land. This step was important because Mississippi law stipulated that control of such rights was a requirement for operating a hazardous-material site. To the opposition, the Indiana University Foundation's move was viewed as a clear sign that the foundation had "become an active participant in the siting of the facility."[24]

Added to the list of reasons for opposition was the apparent unreliability of one of the major contractors associated with construction of the facility, DRE Technologies of Tennessee. This firm was responsible for the design and possible maintenance of the Noxubee incinerator, and its founders were employed by or affiliated with ENSCO of Arkansas. More specifically, "Many worked at an incinerator facility in El Dorado, Ark., which has experienced numerous difficulties including a major explosion in 1989, a fine of $251,633 for emission violation, major hazardous waste spills including one in 1983 of 201,330 gallons, and chronic mishaps with its #2 kiln which burns PCB's."[25]

From June on, the feelings about the siting intensified significantly. Adding to the emotional tone of any exchanges were the Indiana University Foundation's initial unwillingness to meet with student protest leaders, the unwillingness of its president, Curt Simic, to talk to student reporters on the subject, and the apparent beginning of a personal clash between some student protesters and Simic.

During this exchange, the foundation maintained that the foundation had an obligation to maximize the value of gifts to IU, and HESI-FTI's bid of $3.5 million was the highest received since the land went on the market in 1988. They also argued that the best place to determine the use of Mississippi land was through that state's own regulatory and permitting processes, and through the input of its citizens.

Back in Noxubee County

In Noxubee County, protest and support for the facility continued throughout this period. Furthermore, the issues raised by the proposed facility had brought to the surface other long-standing tensions in the community. Although PEON, which remained the most prominent local group to oppose the construction and operation of the facility, was primarily white, another local group, African Americans for Environmental Justice, had also joined the fray.

The community continued to debate the economic benefit versus the potential risks to health and safety posed by the operation of a hazardous-waste facility. There was no question that the promised jobs, if delivered, would be welcomed in a community with few economic alternatives. On the other hand, the potential for damaging the existing agricultural and dairy economy could not be ignored. Protesters and supporters both made frequent trips to plead their cases in Jackson, the state's capital, where their gatherings often had a curious appearance. One state pollution control officer described a meeting in which about two hundred protesters and supporters met in the office of division head Sam Mahry of Mississippi's DEQ: "On the one side of the center aisle were almost entirely white faces, and opponents to the project, and on the other side of the aisle were almost all of the minority people in the audience."[26] The same official remembered the reply of one of the project supporters to an attending Sierra Club representative from New Orleans, who asserted that the decision was an example of environmental racism: "I don't see how you can say this is environmental racism because we are all here, and we are all for it."[27]

Acting as a spokesperson for African American supporters of the project, Brooksville alderman John Bankhead argued that the issue was a power struggle between the have-nots and the haves in the county. He rejected the IU students' charge of environmental racism, countering that the racism was in not allowing African Americans in Noxubee County to make their own decisions and to try to improve their own welfare.[28]

The IUSA was aware of the asymmetric racial composition of the opposing sides but claimed that it did not necessarily reflect the true sentiments of the Noxubee community. The association accused HESI-FTI of using bribery to influence county officials. It further alleged that HESI-FTI committed such deceitful acts as passing out $20 and a free meal to

African Americans, some of whom were not even Noxubee residents, to travel to Jackson to show support for the project.[29]

Some of these African American community supporters, however, argued that the students at IU did not have all the facts. These students, they argued, based their conclusions not on facts but on biased information and a few short trips to Noxubee. They further claimed that if IU pulled out of the land deal based on its determination of "environmental racism" and of what was best for the citizens of Noxubee County, then they would be practicing the worst kind of environmental paternalism. Such paternalism implied that a white institution in the Midwest was more capable of deciding what was good for the citizens of a Mississippi community than the community itself. The argument continued that these same protesters against environmental hazards appeared to have no problem in using powerful and environmentally threatening chemicals in their own enterprises.

Among African American supporters of the project, one of the most vocal and explicit was the local Noxubee chapter of the NAACP. On October 9, 1992, the chapter passed its second resolution in support of the project, in which it explicitly singled out members of the opposition for criticism, and defended its own endorsement of the project on the grounds of the economic benefits it promised to the Noxubee County, especially to its African American community. This resolution, unlike the first, directly raised and rejected the argument of environmental racism.[30]

The IU Foundation Meets with Students

Responding to increased student criticism, and attempting to present a clearer picture of their own position and motivations, the Indiana University Foundation had agreed to a meeting in Bloomington with as many of the key players in the dispute as possible. After a failed attempt to hold the meeting in August, a meeting was finally convened on September 30, 1992, at the Indiana University Foundation headquarters. FTI had originally agreed to attend the meeting but then declined at the last minute. Also canceling their visit were members of the local Noxubee NAACP. The NAACP representatives argued that they had seen campus flyers attacking them in unfair terms and did not feel it worth their while to come. The student associations admitted that flyers were visible around the Bloomington campus, but the students disavowed any responsibility for them. Informally, the representatives felt that they might be going into a hostile environment with the potential for unpleasant confrontation that held little advantage for them.

Attending the meeting were representatives from the Indiana University

Foundation professional staff, the Indiana University Foundation board (the chair of whose Real Estate Committee assumed management of the meeting), the IUSA, an African American state representative, members of the IU Faculty Council Foundation Relations Committee, representatives from the Black Student Union, legal counsel for the university, an assistant school superintendent (who was an African American member of PEON), SEAC, and a Sierra Club advisor. For nearly two hours, while about seventy-five student demonstrators stood and sat outside the foundation entrance, each side stated its position. And although it did not attend, FTI promised to respond to any written questions or comments that student groups wished to submit.[31]

The students at the meeting took the position that it was wrong for the foundation to have given the original options to what was then FTI to buy the Noxubee property, but they realized that short of a major lawsuit (which IU would probably lose), there was no way out of the agreement until the option expired at the end of March 1993. However, the students urged the Indiana University Foundation, now that it had a greater awareness of the situation in Noxubee, not to extend the option past that deadline. The representatives of the foundation promised to review the information they had been provided and to make their decision in due time.

At a closed-door meeting the next month, Indiana University Foundation board members reiterated their contractual obligation to honor their purchase agreement. They did, however, accept a 130-page document prepared by the Noxubee County Committee that directed questions and allegations to FTI on everything from their experience in hazardous-waste management, to the charges of environmental racism, to the health risks of operating a hazardous-waste incinerator. The board left open the possibility that if the option contract expired before HESI-FTI received the necessary permits from the Mississippi DEQ, the Indiana University Foundation was under no obligation to extend the options.[32]

In a curious twist, a number of students involved in the protest argued in November that either the foundation or the university should help pay their expenses. They reasoned that the students were doing the investigative work for the foundation, which had previously been relatively uninformed. The foundation, while not rejecting the request, replied that normally it did not pay expenses without a prior agreement.[33]

After winter break on the Bloomington campus, students again stepped up their protesting, including a four-day outdoor vigil in late January 1993. The vigil was capped off by about a hundred students marching into or around the administrative office of the president of the university, and later to his on-campus home, to present him with a list of concerns and questions. President Ehrlich was not on campus at the time but later replied that when making his decisions on the matter, he would consider all information provided.

Protesters turned to Indiana political sources during this same period. A request for a resolution condemning the Indiana University Foundation land sale was submitted to the Bloomington City Council by the Noxubee County Committee. The council also agreed to consider the information provided by the students.[34] Although some predicted that a quick formal consideration of the issue was possible by the council, no formal action was ever taken, although discussion and debate did occur at a later meeting. Likewise, a powerful African American member of the Indiana House of Representatives, William A. Crawford, who in the past had shown a willingness to look at charges of racism at the university, was asked to investigate the sale. In his reply to the foundation, he revealed that on the basis of direct contact with numerous African American state representatives from Mississippi, the project appeared to enjoy the support of the majority of the community.[35]

The Indiana University Foundation
Takes a Stand

At their meeting on February 6, 1993, the Indiana University Foundation board voted unanimously not to renew the HESI-FTI option if it expired after March 31, 1993. The foundation did not indicate what prompted this decision, but students interpreted it as an admission that the original decision was a mistake. HESI-FTI, upon hearing the news, stated that they were going ahead with their application for the Mississippi permits in the hope of gaining them before the March 31 deadline. Mississippi environmental officials actually stated that, given their current investment, even without approval, they might move forward with the purchase: "They've spent in excess of $10 million on the application, so keeping things in perspective, what's another $3 million? They are working diligently on the application. They haven't given us any indication that they're going to drop it."

The official further stated that if HESI-FTI failed to buy the land before March 31, the state would probably stop processing the application.[36]

The Option Expires

As the March 31, 1993, option deadline approached, HESI-FTI was still embroiled in the state permitting process. The company was responding to a list of deficiencies in their original application that the Mississippi DEQ insisted had to be dealt with before permits were issued. Neither HESI-

FTI nor the foundation were commenting on the outcome if the option expired, but expire it did. On April 2, both the Bloomington campus and city newspapers bannered, "No sale in Noxubee."[37] The company had failed to exercise its option to purchase, and Indiana University Foundation officials expressed the opinion that although HESI-FTI had as much right as anyone else to submit an offer to the foundation, the Indiana University Foundation was pursuing alternative purchasers.

Unbeknownst to the Bloomington protesters, on the day after its option expired with the Indiana University Foundation, April 1, HESI-FTI informed the Mississippi DEQ that they now had a one-year option from the new potential owner of the property, a cattle rancher. The next day students demonstrated against the foundation, occupying its entryway briefly. It developed that the Indiana University Foundation had signed a confidential letter of agreement with Thomas Merrill of Salinas, California, on March 17, well before the March 31 option deadline, to sell him the Noxubee Property for $2.5 million—$1 million less than the HESI-FTI offer. Apparently unknown to the foundation, HESI-FTI, aware of Merrill's interest, had contacted him before March 17 to discuss an agreement for the hazardous-waste facility to be built on the property. After the signing of that letter, HESI-FTI again contacted Merrill and on March 31, the deadline for the original option to buy, Merrill and HESI-FTI signed an agreement giving HESI-FTI a one-year option to purchase 500 acres for the waste facility. Merrill still intended to graze cattle on the remaining 5,500 acres of the property.

Students protested that the Indiana University Foundation had misled them and had contrived behind the public's back to ensure that FTI still received their share of the property. They charged that the Indiana University Foundation had already made the Merrill sale on March 17 and thus its announcement that the property was back on the market as of April 1, 1993, was false. In response to these charges of unethical behavior, the foundation appointed a member of the board of directors, John F. Kimberling, to investigate all allegations and promptly return an opinion, which the foundation would then announce to the public. On April 28, Kimberling submitted his report, in which he found no wrongdoing on the part of Indiana University Foundation staff or board members. His report specified that although an agreement had been signed on March 17, because it was contingent on a number of conditions, the deal could not properly be termed completed, and that, in fact, the foundation had continued, even after April 1, to consider other interested parties. He further stated that it should have been obvious that whoever purchased the land from Indiana University Foundation would have been contacted by HESI-FTI and an attempt made at an agreement that would have allowed construction and operation of the hazardous-waste facility.

By April 1993, it appeared that nothing was going to stop the construction of the facility except a snag in the permitting process. Student protesters turned their attention to forming an ethics commission made up of students, faculty, foundation members, and alumni. However, several events moved to change the scenario. In the May elections, most of the elected officials who supported HESI-FTI lost their reelection bids. All of the aldermen supporting HESI-FTI lost their seats, and in Macon, two of the supporting aldermen failed to be reelected.[38] The real brakes on plans were applied by EPA. Carol Browner, the EPA administrator, in May announced tough new policies on toxic waste from incinerators. During the eighteen months that followed, no permits would be issued for new incinerators, and all existing temporary permits would be reviewed. These policies halted the Noxubee project for the immediate future. The project was further delayed by a Mississippi court ruling by Judge James E. Graves that the state environmental permitting process had not followed all the rules for open hearings and that major portions of the hearing process would have to be repeated under a different structure. Specifically, "the state environment regulators were expressly prohibited from processing hazardous-waste treatment facilities permit applications or evaluating these permit applications." If it did not appeal this decision, the state would have to begin again its hearings on a capacity assurance plan, which included the Noxubee issue.[39] Then Governor Kirkwood Fordice unsuccessfully attempted to overturn Judge Graves's ruling. The end result was no viable chance for a permit until 1995. By that later date, most of the people supporting the facilities had left Noxubee. Although several subsequent attempts were made to establish other waste facilities in other parts of Noxubee, by the end of the 1990s, none had been successful.

In a further development, the Sierra Club Legal Defense Fund on September 2, 1993, submitted to the United States Commission on Civil Rights a brief arguing that the Mississippi permitting process that was allowing the Noxubee County facility siting was a violation of Title VI of the 1964 Civil Rights Act. The Sierra Club requested the commission to investigate these charges and bring a recommendation to the president and Congress requiring Mississippi to carry out its plans for hazardous-waste disposal in a less discriminatory fashion.

Follow-up

In 1996 and 1997, the Indiana University Foundation launched a major fund drive with a goal of $350 million. By early 1998, the pledged contributions had exceeded target amounts. There was no indication that the Noxubee County incident had any effect on the foundation's fund-raising abilities.

INDIANA UNIVERSITY STUDENT ASSOCIATION RESOLUTION

OPPOSING THE SALE OF LAND IN NOXUBEE COUNTY TO FEDERATED TECHNOLOGIES, INC.

Filed May 26, 1992, Considered on June 6, 1992, Final Disposition 21-0-0

WHEREAS, The Indiana University Student Association strives to present and support student ideals as well as uphold the principles and standards of the Indiana University Community, and;

WHEREAS, 6,000 acres of land in Noxubee County, Mississippi, was committed to the IU Foundation for Indiana University by Martin and Opal Conrad in 1975 with the understanding that the land would be sold at an appropriate time, and;

WHEREAS, The IU Foundation has leased the land with an option to buy to Federated Technologies of Mississippi, Inc., who intends to use 400 acres of the 6,000 acres for a hazardous waste conversion and disposal facility and an incinerator; and the remaining 5,600 or more acres would be used for an animal and hunting preserve, residential and industrial development and agricultural use, and;

WHEREAS, IUSA feels that Noxubee County has been chosen as a "sacrifice zone" due to its demographics which include a ⅔ African American population, 42% of the adults of which have less than an eighth-grade education, many of whom are functionally illiterate, a county ranking of 74th out of 82 in per capita income in Mississippi, and a 14.7 percent unemployment rate in June, 1991, and;

WHEREAS, According to Southwest Research and Information Center,
  "No one segment of society should have a monopoly on clean air, clean water, or a clean workplace. Nevertheless, some individuals, neighborhoods. and communities are forced to bear the brunt of the nation's pollution problem. People of color are disproportionately affected by industrial toxins, dirty air and drinking water, and the location of municipal landfills, and hazardous waste treatment, storage and disposal facilities. Environmental inequities are created and maintained by institutional policies that favor whites over people of color."
The policy of exploiting African Americans, Native Americans, Hispanics, and other minority communities by using their land for treatment or dumping of hazardous waste is now more commonly known as "Environmental Racism," and;

WHEREAS, Federated Technologies has been offering a false pretense of economic growth and prosperity to the people of Noxubee County by putting money into the community and promising jobs to people, and;

WHEREAS, FTI has distributed 700 applications while only 100 jobs will exist and, in general, employment at waste management centers is highly specific except for short-term construction jobs, and, therefore, the people of Noxubee County will not truly benefit from the incinerator, and;

WHEREAS, in the majority of incinerators or proposed incinerators, the employment rate has actually gone down because some businesses leave and new ones will not come (source Greenpeace Video: *Rush to Burn*), and the property value of land is $33,000 less than the U.S. Average in communities with proposed incinerators and $36,000 less in communities with existing incinerators according to the U.S. Census data from 1980, and;

WHEREAS, According to information from Noxubee County, the land in question is being sold at an inflated rate without concern for the fact that it would hurt the local population and the local economy and would benefit only the IU Foundation, and;

WHEREAS, Incineration is not a solution to hazardous waste because it does not destroy waste but transfers it to another medium. The waste of incineration not only pollutes the air with heavy metals, dioxins, and over 200 chemicals, but also the ground and then the groundwater when the ash needs to be buried, often polluting another site, and;

WHEREAS, Hazardous waste should be addressed through source reduction at the corporations rather than through removal after it has already been produced, and;

WHEREAS, Indiana University Foundation works for the best interest of the University; and the University strives to educate students so that they will be able to make ethical decisions in the future, and;

WHEREAS, From an environmental standpoint an incinerator would be detrimental to the land and people of Mississippi for many generations; and from a social standpoint, the decision to sell the land in Noxubee County to a company that would build a hazardous waste incinerator reflects both a racial and economic bias, and;

WHEREAS, IUSA would encourage the Foundation to make this decision on ethical future-sighted reasons rather than short-term financial reasons.

THEREFORE LET IT BE RESOLVED that IUSA opposes the sale of the land to Federated Technologies, Inc., or any other corporation that intends to use the land for treatment or dumping of hazardous waste, and;

THEREFORE LET IT FURTHER BE RESOLVED that IUSA will actively work to stop the sale of the land by joining a committee developed primarily for this issue to be led by a department of SEAC.

NATIONAL ASSOCIATION FOR THE ADVANCEMENT OF COLORED PEOPLE,
NOXUBEE COUNTY LOCAL BRANCH

RESOLUTION

WHEREAS, it is the purpose of this organization, the Noxubee County Local Branch of the National Association for the Advancement of Colored People ("NAACP"), to protect and promote the interests of the African-American citizens in Noxubee County; and

WHEREAS, we consider allegations of environmental racism to be serious concerns requiring thorough and diligent investigation and, therefore, we have, through our members conducted a thorough and diligent investigation of the methods and practices of companies proposing to build hazardous waste disposal facilities in Noxubee County and have determined that only Hughes Environmental Systems-Federated Technologies Mississippi, Inc. ("HES-FTMI") has fully informed and involved African-American citizens of this county in both the development of its proposed facility and in the role the company should take as a citizen in this community; and

WHEREAS, we are convinced that our determinations are well-founded and generally supported by our membership and their elected representatives on the governing boards of the various incorporated municipalities located in Noxubee County and the county itself as can be seen from the voting on resolutions by each such body to exclusively endorse the HES-FTMI facility wherein all the elected African-American members of such bodies, if any, voted in favor and were critical to the resulting majority votes supporting such resolution; and

WHEREAS, we are convinced, based upon our investigation, that the federal and state regulations governing such facilities are adequate and will be strictly enforced by the Mississippi Department of Environmental Quality in its permitting deliberations and therefore by both it and the Environmental Protection Agency; and

WHEREAS, we have ascertained that the only allegations of "environmental racism" in connection with the activities of HES-FTMI have been promoted by John Gibson, Assistant Superintendent of Education, and Martha Blackwell, spokeswoman for an organization which calls itself Protect the Environment of Noxubee ("PEON") and it is our judgement that neither John Gibson, an unelected county official, nor the PEON group has the best interest of the county, or its African-American citizenry at heart; and

WHEREAS, we have concluded, based upon our investigation of the PEON group, that it was formed and is directed by a group of white business owners who have, acting in concert, attempted to keep African-American citizens in Noxubee County socially and economically oppressed since the passage of the 1964 Civil Rights Act, and before, and has never taken an active role in promoting the advancement or well being or insuring the environmental safety of African-American citizens, as can be seen in the fact that some members of the PEON group have, in their businesses, exposed African-American citizens to environmental hazards with no regard of their well being, in the group's failure to oppose the location of the Noxubee County Solid Waste landfill or its operation therefore, as well as in the group's failure to contest the facility proposed by United States Pollution Control Industries of Mississippi ("USPCI"); and

WHEREAS, we have concluded, based upon this investigation and our perception of recent events, that a much more serious form of racism exists in the paternalistic attitude exhibited by the PEON group and white business owners of Noxubee County; and

WHEREAS, we have concluded, based upon our investigation of USPCI and its activities, that USPCI has not involved to any extent nor fully informed the African-American community in this county of its proposal to build a hazardous waste facility, and to the contrary, has taken steps to avoid, and thereby disempower our community, as can be seem by its efforts to promote, by contributing funding to pay for the services of an annexation consultant, the annexation of the proposed site of USPCI's hazardous waste facility by the municipality of Shuqualak, the only municipality within the county without an African-American on its governing body; and

WHEREAS, we have concluded, based upon our investigation of all the facts and circumstances surrounding these matters, that HES-FTMI has demonstrated responsiveness to the interests and concerns of African-Americans in this county which suggests a high likelihood that it would protect and be responsive to the concerns, but not limited to environmental concerns, of African-Americans in its future decisions as a corporate citizen in this community, that the project proposed by HES-FTMI would be most likely to enhance the economic well being of African-Americans in this county as it anticipates a larger investment and more employment opportunities than other competing proposals, that the project proposed by HES-FTMI is least likely to do injury to this community or its African-American citizens, and that the project proposed by USPCI would be likely to continue and be an instrument of the traditional oppression of African-American People within Noxubee County;

BE IT THEREFORE HEREBY RESOLVED that we, the Noxubee County Local Branch of the Nation Association for the Advancement of Colored People, declare the issue of environmental racism to be inappropriate and have no connection with the location of a Research and Development facility, Advanced Technology facility, and Hazardous Waste Disposal Facility to be constructed and operated by Hughes Environmental Systems–Federated Technologies Mississippi, Inc.

RESOLVED, this the 9th day of October, 1992.

Eleven

Policy Directions and Recommendations

No matter what form the first major environmental-justice policies take—however well or poorly they are crafted, intended, or supported—they will set the tone for the discussion that will surround all subsequent government activities. Although most policy mistakes can later be rectified, the process of policy legislation will proceed much more smoothly if the enacted regulations do not contain too many wrong assumptions and misdirected initiatives.

If you assume that the issue of the legitimacy of government involvement in particular environmental-justice problems has already been resolved, the question then arises, What do you do next? In the last few chapters, we have explored policy justification as well as problem measurement and analysis. We now move into the realm of policy agenda and policy formulation.

This chapter will not provide detailed policy recommendations for the many types of environmental-justice problems in this country and abroad. It would be both intellectual audacity and poor policy analysis to now begin prescribing specific policies to remedy the ills of environmental justice. Ultimately, specific solutions for the huge category of geographically specific problems, such as facility location decisions, will have to be fashioned locally, although within a national framework. For more general cases, such as pesticide poisoning, detailed policy recommendations cannot be made without an accumulation of information beyond the scope of this book.

What this chapter will offer are a few recommendations and observations on the directions such policies might take and in some cases have already taken. But first, a caveat: in the area of environmental justice, all-encompassing, inflexible public policies and remedies will not work. At one extreme, policies, such as the cessation of all toxic-waste production and a complete halt to their disposal in poor or minority communities, are simplistic and unworkable. They have no practical policy value other than as interesting (and debatable) statements of long-term social goals. At the other extreme, equally unacceptable is the pronouncement that there is no environmental-justice problem, and thus no need for any extraordinary efforts in this area. Such declarations imply that any observed differences in environmental impact are merely the result of market forces in action and that the government should leave well enough alone.

The first grand scheme presumes a miracle, and the second presumes the absurd. The first policy position—unconditional cessation—ignores the hard realities of current society. The production of hazardous waste, the use of chemical pesticides, and a myriad of other environmentally dangerous activities will not soon cease. The environmental policies and practices of today must not, therefore, start from a presumption of absolute prevention. Realistic policies must address what to do when these real-life hazards occur, and how to manage their occurrence.

The second recommendation position seeks, with one grand assessment, to make the entire issue of environmental justice into a simple exercise in rhetoric: There is no problem, just people unhappy with the legitimate and "natural" operation of unbiased market forces. Proponents of this position would have us believe that in our society, with its long history of pervasive race, ethnic, and class bias, a large and diverse sphere of economic and social activity has somehow entirely escaped this poisoning. For supporters of this policy position, there are no wrongs to remedy or rights to redeem. But the sad fact remains that from education to military conduct, from housing to entertainment, from marriage and other social contracts to jobs and other economic contracts, racism and prejudice have been, since the founding of this country, an integral part of its social, political, and economic fabric.

As noted several times here, social issues move through several evolutionary stages. The successful ones survive and progress all the way to the formulation, and then implementation, of policy, before ultimately becoming embedded in our socioeconomic makeup. Although this description is a slight simplification of the development of the issue of environmental justice, the major outline holds. Environmental justice possesses most of the characteristics necessary for incorporation into the basic policy agenda of this country. Its time has come. We must now explore which policies should be included in the environmental-justice agenda.

Of course, much of the basic form of that agenda has already been shaped by the nature of the earlier stages of the movement. This includes its virtual isolation from the mainstream environmental movement. For example, the environmental-justice movement remains a grassroots movement based in local communities, primarily populated by the politically disenfranchised. Furthermore, the movement calls upon its membership for active and direct participation in the pursuit of solutions to issues. None of these features characterizes most mainstream environmental groups. The majority of these mainstream groups are predominantly white, middle to upper class, and not based in the community. These mainstream groups at the national level also rely primarily on their membership only for financial support.[1] Any successful agenda in environmental justice must reflect those realities of organization and membership. This attitude is reflected quite well in the seventeen Principles of Environmental Justice adopted at the First People of Color Environmental Leadership Summit, in Washington, D.C., October 27, 1991. The preamble to this document reads:

WE, THE PEOPLE OF COLOR, gathered together at this multinational People of Color Environmental Leadership Summit to begin to build a national and international movement of people of color to fight the destruction and taking of our lands and communities, do hereby re-establish our spiritual interdependence to the sacredness of our Mother Earth; to respect and celebrate each of our cultures, languages and beliefs about the natural world and our roles in healing ourselves; to ensure environmental justice; to promote economic alternatives which would contribute to the development of environmentally safe livelihoods; and to secure our political, economic and cultural liberation that has been denied for over 500 years of colonization and oppression, resulting in the poisoning of our communities and land and the genocide of our people, do affirm and adopt these Principles of Environmental Justice.

Guiding Principles

The principles that follow the preamble recognize that the environmental-justice policy process is dynamic; it will evolve and even change focus as its major elements change. Today, the primary focus of much of the movement is the siting of local facilities and their operation. Ten years from now, the focus could be on international disposal issues or environmental health care. Beyond simple shifts in underlying economic forces, concepts such as environmental justice, equity, and equality are not absolutes. As our sense of these concepts evolves, so may the underlying goals we attach to them.

The general principles that follow are thus reflections of these concepts as they are currently perceived. They are intended as working principles, not as truths chiseled in stone.

From this perspective, the following should serve as the chief goals of any current environmental-justice policies at any level of government:

- Such policies must promote conditions within a community, group, or workplace that will provide opportunities for fair participation by all parties affected by environmental policies or activities.[2]
- Such policies must address the remediation, compensation, and redistribution of inequities in burdens and benefits.
- Such policies must avoid forcing a paternalistic outcome on the affected parties. To do so would be to replace one form of injustice with another.

Principles of National Policy

To have even a remote chance of affecting the problem of environmental justice, policy initiatives must be concurrently brought forth at several levels of government and within several nongovernmental organizations. Besides a national approach to environmental-justice policy, both state and local initiatives will be necessary to carry out the details of many environmental-justice solutions. Large classes of environmental-justice problems require a flexibility and a familiarity with the specifics of particular problems that can only be achieved at a level of primary operation below the federal government. As the history of environmental regulation in this country clearly shows, broad, one-size-fits-all national policies and detailed national regulations do not work well at the local level, where some of the problems must be solved. Local councils, local community groups, education on environmental policies and risks, and local remedies appear the most promising avenues of solution for many of these problems.

At the same time, these policy solutions, regardless of level, must address not only issues of how the business of environmental activity is conducted, but—just as important—these solutions must take into consideration that different levels—federal, state, regional, and local—may be mixed in the execution of any comprehensive policy plans. Furthermore, this execution will occur at levels and with parties where and with whom the issue of the social impact of environmental policies has previously not been of primary interest.

This consideration also includes a necessary redefinition of just what an environmental activity is and who has policy standing in such issues. Standing refers to stakeholders in an environmental-justice problem. Many stakeholders, both in and out of government, may not even be aware of their standing. For example, the involvement of federal, state, and local agencies—such as the military and many social service agencies—in environmental activities often goes unappreciated, and yet they must deal with the environmental consequences of those activities. Thus, a necessary component of any policy response must include focused education of parties outside the traditional environmental governmental and nongovernmental communities.

To return to an earlier assertion, why is any national policy necessary? As should be evident by now, regardless of where one stands on the issues, most location- or geographic-specific environmental-justice problems cannot be depicted with a single, broad federal brush. The combination of players, circumstances, and conditions in such a locally specific problem often requires a local solution. Why become overly concerned about framing a national policy for activities that ultimately have primarily local consequences?

It is true that for many classes of environmental-justice problems, although the details of a solution should be local in nature, a broad federal foundation is still necessary. Why? Put most simply, an overlay of national rules of conduct prevents, or at least reduces, extremes in local outcomes. If no such national policy framework exists, each locality or state could set up entirely different levels of safeguards against inequitable environmental decisions and activities. Any rational firm—all other conditions being equal—would obviously select locations with the least restrictive, least costly regulations. In today's environment, that least costly location is usually the one that offers the least political resistance and the lowest-priced land.

A national structure for environmental-justice treatment would provide for both geographic-specific and, where possible, for population-specific environmental-justice problems—a foundation on which local policy decisions could be made. A major challenge will be providing such national parameters without restricting the ability of local decision-makers to respond in an effective manner to the specifics of the environmental-justice problems they face. Local governments have always had primacy in determining community land use. Thus, for those environmental-justice issues involving local land use, any national structure would be superimposed on an existing local regulatory network. And it is not clear just what the reaction of local community governing bodies would be to such national intrusion.

Nevertheless, since fall 1994, several bills have been repeatedly introduced in Congress that address various aspects of the environmental-justice issue. Some have called for the establishment of new offices and new mandates within the U.S. Environmental Protection Agency (EPA), and others have proposed a broader, less specific mandate for the environmental and natural resource policies of many federal agencies that may have differentiated socioeconomic and health impacts on various populations. By the beginning of the new century, none of these particular bills has been enacted, and none is likely to be enacted soon.[3] Assuming a continued maturation of the environmental-justice agenda, regardless of the specific party in control of Congress, in time, some federal legislative action will likely succeed. However, given the legislative history of other social issues, including the original environmental agenda, the enactment of one piece of legislation will almost certainty not provide environmental justice for all.

While neither supporting nor criticizing any particular piece of current legislation, at the national level, certain procedures must be avoided.

- Avoid adopting any policy that seeks to achieve environmental justice via a series of "command-and-control" regulations. In most areas, but especially environmental and natural resources, the federal government's history with such regulations clearly demonstrates that such an approach lacks both process and cost flexibility, and furthermore, often fails to achieve the desired outcome.[4]
- In defining minority or lower-income status, avoid fixed-percentage definitions. Such definitions often take the form, "Minority communities or minority labor forces are those with greater than x percent minority." Favor instead relative-status definitions, wherein status is based on the difference in minority percentage between a given community and a control community. Or make comparisons along a continuum: compare changes in some condition or status as the percent of a target group changes.
- In seeking to correct historic conditions, avoid making some determination of villainy the prime focus of the regulation or policy, or a precondition for any subsequent remediation activities. Granted, in some cases, clear identification of and sanctions against past violators can be part of a remediation process. Nevertheless, avoid environmental policies that hinge on such assignments. In many cases, responsibility is defused and confused. Valuable time and resources, better expended on corrective actions, are wasted in such pursuits.
- Include in any national policy initiative or regulatory plan explicit provisions for the establishment and enforcement of education for the local population and their participation in the decision-making process. Local communities must be given some real influence in decisions that affect them. Without such participation provisions, environmental exploitation could be replaced by environmental paternalism.

Given local community participation in the final outcome process, variations in outcomes across communities should be expected and permitted. A major criticism of much current environmental regulatory policy is its rigidity, its inability to accommodate local realities.

A cautionary note: this idea may sound reasonable, but if it is not carefully structured, it holds within itself the danger of placing local communities at the mercy of whoever yells the loudest. Nevertheless, community self-determination implies acceptance of the risk that outcomes may not always meet expectations.

- Avoid trying to use the resolution of explicit environmental-justice issues as a device for solving far larger social problems. For example, a community, lacking other economic opportunities, may decide to assume a higher level of environmental risk than the population as a whole in exchange for some compensation. Some environmental-justice advocates have called this situation environmental economic blackmail. These advocates correctly identify the lack of economic opportunity as the reason for a community's willingness to assume the possibly increased risk. Where the policy problem occurs is when such advocates argue for prohibiting such arrangements. Denying communities a right of self-determination on the grounds of preventing "economic exploitation" is both paternalistic and bad policy.

It should not be the goal of environmental-justice policy to overcome all the economic inequities in society in a single gesture. Environmental-justice policies may help to set a level playing field in areas such as negotiations on environmental activity outcomes to avoid "unfair" bargaining strategies. But such policies should not seek to prevent all participants in such a bargaining process, when properly informed, from negotiating their own settlements.

In all these policy formulations, there should also be a frequent and serious evaluation of the changing current conditions. For example, it is common for environmental-justice proposals on hazardous-waste disposal to recommend regulation that effectively prevents such disposal in minority or poor communities. To be sure, prevention should in nearly every case be the preferred long-term policy goal. But the reality of today, and likely for the foreseeable future, is that many forms of hazardous waste are still being generated and must still be disposed of. As a current social policy, NIMBY ("not in my backyard") cannot succeed. Waste must go into someone's backyard. Environmental-justice solutions that ignore this hard fact have an intrinsic flaw. The solutions must accept the fact that environmental burdens cannot be made to vanish totally. They may be reduced at the aggregate level. That is, the sum of all environmentally hazardous waste that has to be processed may be reduced. They may be more evenly distributed across various groups and locations. Also, there may be more equitable compensation for enduring such burdens. But the burdens cannot be completely eliminated. Someone's backyard will get them. The issue is not if, but rather who, how, why, and for how much.

The guideline for the substance of national policies can be taken from the earlier discussion on policy analysis: Why should government get involved? Market failure, although not the only reason, provides a powerful rationale for government intervention. A major, if not the dominant, impetus for any national environmental-justice agenda must be the remediation of those circumstances in which the market mechanism—without government intervention—results in an inefficient outcome. Again we are reminded that this market-based remediation does not negate government action that addresses issues of human dignity or some principle of equity. What it does say is that market failure interventions are much easier to justify on the grounds of actually improving rather than distorting the operation of the economy. This position appreciates that historically, some governmental interventions (for example, municipal rent control or the liability assignment provisions of EPA's Superfund) have actually in many cases worsened market failure and contributed to a worsening of economic operations.

Many of the equity issues in the environmental-justice debate may also be issues of market failure. It is an equity issue when farm workers endure uncompensated risk due to high levels of pesticide exposure. It is also a market failure because the farm workers are not compensated for the negative health risk externalities generated by the farmer. It is an equity issue when a hazardous materials facility locates in a particular community because of information asymmetry that favors the firm over the community. It is also a market failure because, just as in the case of the pesticide-using farmer, the facility is not paying the true cost of its production but instead is paying a lower amount, subsidized by the uncompensated local population.

Another policy approach note: many analysts, advocates, and practitioners of environmental justice seem to still be fixated on proving that the more powerful force determining the location of hazardous-material or landfill sites is class or income, rather than race or ethnicity. This is a spurious debate. At several levels, it is irrelevant. It is likely that in many community situations, class or income is the more powerful determining factor, and it is likely equally true that in others, the more powerful force is race. Both explanations involve issues of social impact; both suggest a rethinking of the mechanics of siting. And in either case, the possible remedies are virtually identical. Furthermore, it may be assumed that in yet other instances, the location decision may truly be random. A reasonable consideration of even the sparse information now available suggests that no single mechanism for location decisions operates across all environments. Part of the national objectives of environmental-justice policy should be to avoid such tertiary debates, which have little bearing on the directions and nature of the ultimate solutions.

The role of the federal government in providing a broad-based foundation for the solution of problems of environmental justice may take the following forms:

- Providing baseline criteria for compensating a community or population for unequal environmental burdens. More than any other national responsibility, this would help to avoid the problem of firms or enterprises "shopping" the states for the least inconvenient or least costly location. Without such guidelines, interstate variation would simply mean a transposition of the terms of inequity from smaller groups to the states.

- Recognizing the potential problem of "intergenerational equity" as a significant part of this compensation baseline. Benefits and burdens do not occur in an instant in time but over an extended period. In many cases, that period lasts so long that those making the initial decisions and receiving the initial compensation may be out of the picture before the major aspects of the environmental burden are realized. If environmental effects are intergenerational, so should be compensation or remediation.[5]

- Devising baseline criteria for involving all communities or populations in the policy-making process. This may include guidelines for a community's response to siting or remediation decisions.

- Serving as the primary provider of key information on environmental activity. Most important is the development of systems for integrating such environmental information with relevant demographic information. This also means reexamining the overall process of risk measurement at the national level to include measurement that more accurately reflects the health of entire communities or populations. For example, data on cross-media, air and ground contamination, and cumulative risks must be collected, along with information on noncancerous, nonlethal impairments to health, such as asthma or reduction in cognitive abilities.

- Providing mechanisms for correcting the information asymmetry problem. This means more than assuring that demographic and environmental information for the community is available, although this information obviously should be provided. The information must also be made accessible to community members in a useful form.[6]

- Determining the basic rules of accountability and compensation for historic inequities. This responsibility could include mechanisms for who pays what, and for how long. As noted earlier, overemphasis on the assignment of past "guilt"—besides opening up a difficult series of legal battles—would take attention away from the more primary need to alleviate such problems.

In a later section of this chapter, I will examine specific steps the EPA has taken after the 1994 Executive Order 12898, which address some of the policy concerns expressed above.

Reexamining the Old: Applying Existing Regulations More Fairly

As both a political practicality and a policy strategy, staking the success of the national environmental-justice movement on the successful passage of new legislation invites delay and disappointment. Although the environmental-justice movement has definitely gained both recognition and power within the last several years, neither its current reputation nor its power are sufficient to bring about a major set of legislative pieces. Success in the immediate future will depend less on the creation and passage of new legislation than on the reexamination, interpretation, and execution of existing legislation.

But some may note that the major federal environmental laws and regulations—such as the Resource Conservation and Recovery Act; the National Environmental Protection Act; the Comprehensive Environmental Response, Compensation, and Liability Act; the Clean Air and Clean Water Acts; the Toxic Substances Control Act; and the Federal Insecticide, Fungicide, and Rodenticide Act—do not contain provisions that explicitly address the issue of unequal impact of environmental activities. However, all these acts do have provisions or declarations of intent that promise protection or relief for "all Americans" or for "each citizen" (or similar language signifying that everyone has a stake in the act's mandate).

At the national level, any environmental-justice policy initiative must include a systematic reexamination of the administration of such acts, with an eye to assuring that all segments of the population benefit equally from the acts. The spirit of this policy initiative has already begun with President Bill Clinton's Executive Order 12898 on environmental justice, which directs all federal agencies to review existing regulations and procedures in order to identify any inequities. The review process must disaggregate the various activities into assessable parts in order to answer the following questions:

- By use of several different measurement instruments and at different scales of analysis,[7] do the individual components of the act result in an uneven distribution of environmental benefits or burdens—including health, economic, and social outcomes—across demographic or regional groupings?

- If such unevenness exists,[8] can it be explained by neutral factors beyond the control or responsibility of the act or agency?[9]
- As part of any impact analysis of existing regulation, are provisions for information or citizen assistance structured in such a manner as to handicap subsets of the population?[10]
- In reviewing agency policies, are operational and administrative procedures—even if they have not yet been identified as definitely leading to biased outcomes—conducted in a manner that disadvantages certain groups?

With sufficient inspection, almost any act under almost any circumstance could be found to have some distorted outcomes or procedural inequities. And all these problems cannot, perhaps even should not, be fixed. The complexity of modern society almost guarantees that no law can be truly impact-neutral. The issue, therefore, is one of intent and degree. Application of the concepts and principles of environmental justice to existing laws and regulations does not mean a witch hunt. It does mean that a careful and ongoing analysis of the content and administration of existing laws can, from the perspective of equity, improve the impact these laws have on the total population.

Federal Environmental-Justice Actions, Regulations, and Responses

As noted earlier in this chapter, although there had been some broad-based environmental-justice legislation proposed during the early 1990s, there has not been much federal legislative action or prospect for successful future action by the turn of the new century. Reflecting this short-term reality, Presidential Executive Order 12898 in 1994 was one attempt at arriving at some national environmental-justice strategy and policy without relying on any new legislative mandate.

Executive Order 12898 mandated all involved federal agencies to develop explicit strategies for incorporating environmental-justice concerns into their existing policy and activity processes. Toward that need, an interagency working group was created in early 1994 with the EPA Administrator as chair to guide the strategy development process. Agencies were charged with submitting within a very short time an environmental strategy report. What made this interesting was that in most cases, these agencies had, before the executive order, likely never given a thought to the concept of environmental justice, much less actually considered how these concerns might impact their agency. But informed or not, within two years of the executive order, a number of agencies, such as the U.S. Navy, Housing and Urban Development, and the Departments of Transportation, Energy, and Interior, developed and published such environmental-justice strategy pieces. Without exception, these strategies essentially involved a statement of "commitment to the principles of environmental justice" by the agency and a listing of ways in which information about environmental-justice concerns would be incorporated into parts of the agency policy or activity process. In no case were explicit agency activities tied to satisfying environmental-justice concerns, nor was environmental justice incorporated into the working fabric of policy sections of the agency. Incorporation means having an office of environmental justice or equity as part of a central policy group within the agency, instead of the almost universal practice in agencies, as outlined in their strategy reports, of placing environmental-justice activities and services in an information or public affairs subgroup.

However, this passive reaction to Executive Order 12898 was not the case within the EPA. Whether recognizing the futility of direct legislative action or simply recognizing the inherent possibilities that always existed under current statutes (or recognizing their inherent environmental-justice responsibilities), the EPA put considerable effort into addressing environmental-justice concerns through regulatory reform. The EPA came out in April 1995, a little over a year after the original executive order of February 1994, with their environmental-justice strategy report. In a manner similar to those of other agencies, the report devotes considerable time to information and dissemination initiatives. But going beyond other agency strategy reports, the EPA explicitly stated its intention to incorporate environmental justice into its regulatory compliance process. Making good on this goal, in the case of the National Environmental Protection Act, the EPA in 1997 issued an interim guidance, and then in April 1998 a final guidance, on environmental justice and the National Environmental Protection Act.[11] The final guidance is interesting both in what it includes and what it does not. The majority of the guidance did identify several general categories of analytic concerns when considering environmental justice, such as defining minority or low-income populations, and establishing methods for considering the effects of population characteristics, geographic factors, economic factors, and health risk factors. The guidance also directly confronted the problem of community participation and how traditional EPA public participation methods and approaches may work against adequate participation of some minority and low-income affected populations.[12] However, the guidance did not mandate any particular approach or method of analysis. Perhaps recognizing the primitive state of environmental-justice methodology and the myriad conditions gathered under the environmental-justice label, the EPA only outlined dimensions of concern and suggested methods for addressing problems.

A far more controversial EPA environmental-justice initiative has been its guidance on Title VI of the 1964 Civil Rights Act and the process of issuing pollution control permits. The interim guidance on Title VI was issued in February 1998.[13] Its focus is the identification of discrimination in the permitting process. Discrimination is defined as occurring when the permit would cause a "disparate impact" on a racial or ethnic population without an "acceptable mitigation plan."[14] As the new century begins, several cases are pending under this guidance, which a wide spectrum of big-city mayors, industry representatives, and state, local, and federal lawmakers from both parties have opposed generally on the grounds of causing job losses and driving needed industry away.

The EPA's reaction to this criticism was neither to make major modification in the guidance, nor to withdraw it. Approaching various stakeholders' concerns, the EPA sought to alleviate their fears. For example, a major component of the recent EPA initiatives in economic development and environmental justice has been brownfields redevelopment, more particularly the Brownfields Initiative Pilot Projects.[15] The argument is that needed economic development and community improvements can be realized by effective reuse of land or areas that have been previously held under a cloud of environmental site contamination or hazardous-material location. In this program, the EPA provides grants for pilot project start-up, which includes encouraging public–private partnerships and community outreach. Environmental-justice efforts were placed as a key evaluation criterion for the pilot projects. Several critics of the guidance on Title VI have argued that the Title VI regulatory initiative in pollution control permitting would impede needed brownfields redevelopment. This in turn would harm the economic well-being of many urban communities.

In part as a response to such criticism, the EPA launched a series of case studies of the brownfields pilot.[16] A panel of brownfields stakeholders, most of whom were outside the EPA, were charged with investigating a total of seven brownfields pilot projects in order to assess the impact of Title VI requirements on pilot operation. It was found that although there was certainly environmental-justice activism in the communities, there appeared to be little desire for filing Title VI complaints. This in part may reflect the mitigation inherent in the environmental-justice component of the original brownfields grant or inherent in the sense of existing community involvement in the brownfields decision-making process.

Of course, the brownfields case studies result does not answer the criticism that Title VI impedes economic development. Many permitting situations are not brownfields, and in fact, as noted earlier, the required environmental-justice component clearly positions brownfields projects to avoid community complaints. Nevertheless, the concerns that Title VI hinders economic development may be overstated.

Our International Obligations

Largely neglected in most of the discussion have been this country's international responsibilities toward environmental justice. From the standpoint of both policy and politics, it would be a curious ethical solution if, in order to achieve major improvements in the environmental-justice conditions of this country, we simply exported our inequities. To cite a particularly pertinent example, the problems with the disposal of hazardous materials should not be solved by simply dumping them in another country.[17]

Very much the same economic forces that make minority and low-income communities attractive locations for hazardous-material facilities also operate abroad. Developing and recently developed countries often exhibit the same dire economic need, susceptibility to expedience, low land values, and information asymmetry as do minority and low-income communities in this country. If anything, the severity of both the need and the distortion of information is often greater for a developing country, in which there are few perceived economic alternatives and a poorly educated and poverty-stricken population. For example, Haiti, the poorest country in the Western Hemisphere, with one of its least educated populations, has already become the favorite dumping ground for the toxic waste of other countries.[18]

In a country whose population has a relatively low life expectancy, long-term threats from exposure to toxic substances lose much of their impact. A carcinogen that may kill you at the age of sixty-five does not seem as threatening when the average life expectancy is only fifty-five years. The primary worry in such a country is finding adequate food and shelter for next week.

The entire topic of international environmental justice is far too complex and multidimensional to adequately treat in this book. Issues such as the international trade-offs between environmental risk and economic compensation should not be ignored. However, in focusing primarily on domestic issues of environmental justice, these just as potentially critical international issues will not be treated here.

Defining the Community: The Correspondence Principle

Returning from the international to the focused local community level, among the many other decision-making issues faced at that local level is that of determining who has the primary authority for making decisions that affect individual communities. Currently, in too many cases, the local level at which such decisions are made is considerably higher than the level at which the impact of those decisions is felt. County or city planning commissions, for instance, make decisions that affect much smaller and often much darker or poorer neighborhoods. In Houston, the majority white government has allowed, if not encouraged, most landfills and hazardous-material facilities to be located in minority neighborhoods.[19] In Chicago, the city and county governments have permitted the highest concentrations of hazardous facilities in the poorest minority neighborhoods, even over neighborhood objections.[20] As Robert Bullard (among others) has pointed out, local decision-making processes simply negate or minimize the objectives of the smaller minority neighborhoods, where these decisions will have the greatest impact.[21]

Inherent in any local decision-making process is the problem of how to incorporate proper representation. Many location decisions have failed to do so. From the location of highways that divide and pollute minority or poor communities to the placement of older city dumps, the undesirable are usually among the unrepresented. With at-large elections and other electoral mechanisms that result in a weakening of the political voices of smaller neighborhoods, a county or city body elected by the general population cannot be relied on to serve the interests of smaller neighborhood groups.

The denial of representation to affected neighborhoods, even more than the asymmetry of information, is the most serious problem of environmental justice facing individual communities. This lack of influence in the decision-making process is one of the primary reasons some have so strenuously argued for a national regulatory administration for environmental justice and a nationally imposed solution. But policy and regulation governing land use has been primarily a local prerogative.[22] Although the environmental-justice movement has gained much prominence over the last several years, there is no evidence to suggest that it will somehow radically alter this decision-making arrangement.

The problem is not the local nature of decisions that determine the location or operation of facilities, but rather the mismatch between those making the decisions and those enduring the consequences of those decisions. When the decision-makers do not bear the full cost or responsibilities for those decisions, there is a distortion in the decision outcomes. They do not bear the full cost of their decisions because only a fraction of the city or larger community actually has the environmental risk located near them. The larger body makes the decisions, but a smaller subset bears most of the cost. This in some ways is similar to the old saying that "old men make wars, but young men fight them."

Given the short range of the environmental effect of many facilities, it is possible that the citizens only a few miles from a site face virtually no health or economic costs from permitting such a siting. And assuming that the affected neighborhoods have little political influence on the decision-making body in the larger community, the larger community can enjoy the economic benefits of the site at a low cost. Members of such a larger community will thus be tempted occasionally to approve potentially harmful sitings.

This problem of mismatch between benefit recipients and cost bearers is, of course, not limited to the environmental-justice area. Virtually every decision on local zoning or economic development is complicated by elements of this problem. In public finance, this problem even has a name: the correspondence principle.[23] The correspondence principle argues that the most efficient system for the delivery of services or benefits will ensure that the jurisdictional unit making the key decision or having the most authority should correspond to the unit bearing the primary costs for the benefit. Benefit spillover is what occurs when the two are mismatched.

As noted at the beginning of this chapter, it would be presumptuous to make a long list of specific policy remedies. But a definite policy principle does emerge here. Whether implemented via a nationally legislated mandate (very unlikely), a provision legislated by the state (much more likely), or a locally adopted ordinance or regulation, the correspondence principle should be applied. That is, although not supplanting the decision-making authority of the larger community, the correspondence principle should be ensured via a specific mechanism that those most affected have the loudest voice in the decision-making process.

This amplified voice can take many forms, some of which have a long history in local government. Affected neighborhoods, particularly small ones, can acquire power through neighborhood councils with some legislated authority, through mandated community referendums on key location or operations decisions, or through the creation of special districts. Whatever form is chosen, its authority must be institutionalized and codified. Referendums that depend on the pleasure of the larger community jurisdiction could easily become no referendums. Neighborhood councils with only advisory status and formed at the pleasure of the larger decision-making body become facades of power, with no substance. Different forms of neighborhood involvement and empowerment serve different purposes. The forum for deciding the location of a proposed facility, for example, may not be a useful mechanism for dealing with environmental risks generated by currently operating facilities or with the residuals from previous operations.

Consider community referendum as an empowerment device. Assuming that the affected community has been reasonably defined, the referendum can elicit community attitudes toward major siting decisions. In theory, every member of the community has an opportunity to voice his or her opinion. By redefining for each referendum the dimensions of the affected community, a better correspondence between decision-making and cost spillover can be achieved. Unfortunately, research has demonstrated that as the complexity of the issue increases (past the stage of approval or disapproval), the referendum method does not work as well.[24] Nor does the referendum work well in carrying out the more detailed and continuous decision-making associated with an existing environmental risk—such as managing an abandoned production facility that has become an environmental hazard or dealing with currently operating facilities that have already acquired their location rights.

In many states—New York, for example—special districts are permitted that conform to the boundaries of neighborhoods that receive a certain combination or pattern of municipal services such as water, sewage, or trash removal. Thus, people on adjacent blocks could have different schedules for trash pickup, different levels of water service, different forms of street lighting, and different qualities of roads. Erie County, New York, the location of Buffalo, contains over a hundred such special districts.

Such special districts are distinguished by two key features: they do not necessarily conform to preexisting political boundaries, and residents within these special districts have primary jurisdiction over the services they receive. This form of neighborhood empowerment could be extended to include special districts for location decisions. The special district can also be useful in dealing with existing environmental burdens. Much of the current environmental-justice problem is not preventing new operations but dealing with the facilities that are currently operating, or that have previously operated in a community and have left it with an environmental problem. Local governmental entities already exist, via state charter or mandate. It would set no precedents to permit the creation of such entities.

Similar in form to the special district—or in conjunction with the formation of a special district—would be the neighborhood council, with assigned jurisdictional powers of community environmental management. Within the local, state, and federal regulation of the environment, a host of detailed decisions are carried out at the highest jurisdictional level rather than the one most affected. The community council, or similar organized forms of neighborhood representation, allows ongoing and focused community input into environmental policy issues related to existing structures or historic residuals.

A potential difficulty with both special districts and community councils is that they would still follow fairly rigid geographic boundaries; thus, they might not be able to address an affected area in the way a special referendum could. And there probably is a natural limit to the number of special districts that one community could sustain. Overlapping special districts or councils could create as much chaos as solutions. Nevertheless, because no jurisdictional redefinition or configuration will perfectly satisfy the correspondence principle, for many issues, the special districts and community councils would serve as rough approximations.

Part of the process of ensuring meaningful community representation is a recognition that the definition of neighborhood or community is a fluid one. There is only a slim chance that the affected population would reside within some convenient local voting district or other previously defined jurisdictional boundary. Both the actual location of a particular facility and the specific nature of the environmental burden it poses should contribute to the identification of the "affected" neighborhood. For example, many chemicals likely to be found in a landfill have a radius of impact of less than a mile. Beyond that, their impact will steadily diminish, and the topography of the area can attenuate their effect even further. Conversely, some airborne pollutants can affect people living 5 or 10 miles from their source.

To properly capture these affects, neighborhoods should be defined from the perspective of the environmental hazard, rather than outlined according to some convenient voting district arrangement or similar fixed jurisdictional boundaries. For different specific problems, different definitions of neighborhood may have to be entertained simultaneously. For those committed to neat processes, such flexibility can be disconcerting. Whether the responsibility of neighborhood definition lies at the state or local level, adherence to the correspondence principle, and the associated principle of equitable representation, demands that officials and planners wean themselves off of the obvious attraction of simply defining and empowering neighborhoods along preexisting political boundaries.

Community Access to Environmental Information

Defining affected neighborhoods on a case-by-case basis means that at the local level, the capacity exists for estimating, however crudely, the area of impact for any environmental operation. Such an estimation, of course, requires the availability of toxicological, demographic, and engineering information that may be beyond the often limited human and physical resources of local government or, in some cases, even state governments. As noted earlier, the federal government should not be in the business of micromanaging the specifics of a local solution. However, through such agencies as the EPA or the Departments of Interior, Energy, Agriculture, and Defense, the federal government is the logical source for providing needed technical assistance in defining the impact of specific facilities.

As I noted previously, it is not enough that the information be provided. It must also be provided in a form that is useful to its ultimate consumers. Often the needed combinations of demographic and environmental information are simply not available. But even when they are, these information sets exist in formats that make them barely useful to environmental policy professionals, much less to community members with little formal training.

Perhaps the most important initial objective of the information providers must therefore be the reconfiguration of the information sets into useful forms. This task will be a formidable one. None of the major federal agencies, such as the EPA and Department of the Interior, have long histories of successfully providing policy information directly to the public. The EPA and most of the state environmental protection agencies have never viewed the "public" as primary consumers of their vast sea of data.

At the practical policy level, the need for usable information should not be left to the goodwill of federal or state employees. Providing usable information is perhaps as important a federal responsibility as providing a national standards base for making location decisions. In fact, it is a key element in the national standards base. Any concept of equitable treatment and participation must include a mechanism for providing affected citizens access to the necessary information.

But it is not enough that citizens be able to just walk right in and get a copy of the latest toxic release inventory (TRI) for their state or even their local community. For example, under the Community Right to Know Act, anyone can currently contact the EPA and obtain a disk copy of the latest TRI for their state or download it from the Internet. This information will usually be in either data-based management format, such as D-Base III, or spreadsheet format.[25] But to even view this information, one has to have a microcomputer, have either suitable spreadsheet software or data-based management software, and, most important, be computer literate enough to use the software. So, it is unlikely that the average citizen of an affected area will be able to easily use that information.

And TRI information is just one part of the environmental data picture for a given community. The concerned citizen must pursue similar processes to obtain information on a wide range of environmental conditions — information that they might not even know exists. This information, along with an accompanying array of demographic data, must then be combined by that citizen in order to make an informed decision about environmental activities within their neighborhood.

Clearly, a new system must be developed for the average citizen. About its exact form we can only guess. But it must be easy to obtain and easy to use.

Policy Solutions as Science and as Art

The policy directions and solutions I have presented here run counter to an environmental-justice policy agenda that seeks to replace inequitable systems with unworkable ones. The science of responsible, well-structured policy analysis entails a process of rule-making that relies heavily on systematic, often quantitative, evaluation of each issue. This science rejects a policy approach that is satisfied to simply "put something on the books," however ill-conceived or inadequate.

At the same time, these suggestions do not demand such a strict quantitative approach to problem-solving that an issue is never resolved, but only studied to death. For example, proposals to require solid economic analysis of the impact of all environmental rule-making appears at first glance to be consistent with the recommendations outlined above. However (and this is a big caveat), such analysis should support the rule-making process, not drive it. At present, in many areas of environmental regulation, and definitely within environmental justice, information is simply not reliable and thorough enough to perform such analyses properly. But solutions need not remain in limbo until sufficient information is obtained. For each moment that their plight is not addressed, real people must suffer conditions of inequity.

Ultimately, workable policies for the real world are a combination of science and art. The art lies in realizing that the results of even perfect scientific analysis cannot be realized within the existing limits of time and politics. The art of formulating workable policy depends on the recognition that a part (often a large part) of the policy-formulation process must be carried out under clouds of uncertainty. In spite of the missing pieces, the policy process still moves forward.

The difference between good and bad policy formulation is the degree and manner in which the science and the art of policy analysis are blended together at a point in time to generate a policy product. The good policy process addresses the underlying issues with full and public awareness of what is not known.

Twelve

Environmental Justice
A New Paradigm—A Time of Change

> As much as the environmental dilemma is a problem of ethics
> and epistemology, it is also a problem of discourse. Various
> proposals to resolve the crisis are put forth by different social
> groups with different sources and kinds of information, groups
> with divergent goals, methods, values, and epistemologies.
> —M. J. Killingsworth and J. S. Palmer,
> *Ecospeak: Rhetoric and Environmental Politics in America*

The environmental-justice movement has been portrayed as a movement
about racial injustice. It has also been described as a grassroots movement
that can no longer be ignored. Some critics from the mainstream environmental movement, when the environmental-justice movement first began
gaining attention in the 1980s, had described the movement as a red her-
ring with little if any merit and as a distortion of the founding principles
of the environmental movement. These same critics argued that to succeed
in all that remains to be accomplished, the environmental movement could
not afford to be sidetracked by what they perceived as essentially nonenvironmental issues of social policy. Supporters of environmental justice
countered that it is an issue long overdue, a necessary addition to the environmental agenda. In fact, its supporters agree that environmental justice
does represent a shift from the founding principles of the environmental
movement. But it could be argued that those principles were flawed and a
new paradigm is necessary. These voices argued that without a concern for
environmental justice, environmental policy in this country will remain for
many of its citizens a sham, benefiting some while discriminating against
others on the basis of their race, their income, or their location.

Whether one is a critic, a supporter, or just an interested observer, what
is clear about environmental justice is that it means change for all current
players in the environmental policy game. This change has already begun
at the national, the state, and even the local levels. This change has affected
at the national level not only the U.S. Environmental Protection Agency
(EPA), but many other federal agencies, and has drawn in many of the
national nongovernmental environmental organizations such as the Sierra

Club and Greenpeace.[1] At the state and local levels, there are literally hundreds of environmental-justice grassroots organizations around the country, many of them now with several years of experience and much better prepared for policy actions. Furthermore, at the state and local levels, the idea of environmental justice as part of the environmental policy agenda has definitely made its way into the policy framework of the state and local governmental agencies charged with environmental protection. At this point, it would be safe to say that the environmental-justice movement has moved both nationally and regionally to the point where the issue may still be opposed, but it cannot be ignored.

The EPA, and to a lesser extent other federal agencies, has begun to recognize the paradigm shift which environmental justice represents. State and local agencies have begun to incorporate environmental-justice concerns, language, and issues into their policy-making process. For now, at both the federal and state levels, these incorporations have been executive rather than legislative in origin, but they have nevertheless begun. At all levels of environmental policy-making and activities, the way business is conducted has begun to change. One fundamental part of this change is the slow recognition that environmental policies and regulations affect different populations in different ways; different populations can have different reactions and concerns toward given policies. These governmental agencies are changing.

Mainstream environmental organizations, perhaps at a slower rate, have begun to change and have begun to cease embracing the myth that minorities do not care about the environment. There has begun a recognition that this myth cannot be used as an excuse for the woeful underrepresentation of minorities within the environmental organization leadership and staff. These organizations are changing. In the same way, minority social action organizations can no longer pretend that environmental issues are a concern of only the white middle class. These organizations too are changing. Finally, communities, however defined, need no longer assume that the environmental burden they endure will not be compensated for or mitigated. They need not accept a passive role in the environmental policy decision-making process. Communities are changing.

This change will not always be in a positive direction, but it will occur. There have been and will continue to be ill-conceived and poorly formulated environmental-justice remedies that defy common sense, practicality, and in some cases even the principles of equity of the environmental-justice movement. Furthermore, during this change, the impact of environmental justice as a policy issue at various levels of government and scale will likely continue to wax and wane. Even within the last few years, environmental justice has experienced several changes in attention and urgency at all levels of discussion. This is natural and to be expected. What should not be forgotten is that the genie is out of the bottle—and it will probably never be put back in.

Brief Summary

Throughout this book, I have explored many aspects of the issue of environmental justice—from what it is and how it arose, to how to measure it and what may be done about it. The problem appears simple at first glance, but with each further step of inspection, it grows more complicated. In the end, one is left with both a sense of amazement that the issue took so long to emerge on the national and regional stage, and a sense of despair at ever resolving it. And in fact the goal of environmental equity and social justice will never be achieved, just as the ideal of a perfectly balanced environmental policy can never be realized. But perhaps the most important goal is not actual achievement of perfect environmental equity and justice for all, but rather, as with all other social policy goals, our profession of a willingness as a society to commit resources to keep trying to continually move closer to a constantly redefined ideal.

But environmental justice represents more than an additional set of voices to the dialogue on environmental policy. It even represents more than just a cry for fair treatment from previously disenfranchised communities. Perhaps the greatest significance of the environmental-justice movement, and one I have emphasized in this book, is the recognition that matters of social and economic impact can no longer be ignored when formulating and implementing environmental policy. The entire manner in which everyone approaches questions of environmental policy is beginning to change. The old way, which focused almost exclusively on the narrowly defined physical consequences of a particular environmental event, has less and less relevance in a society in which the socioeconomic consequences of environmental decisions can often overshadow the attention to details about water purity, soil contamination, and ambient air quality.

The most vexing problems facing environmental policy decision-makers at the beginning of the twenty-first century is not whether a given production process produces a given toxic substance, or even how well that substance can be regulated. The central problem at all levels of decision-making will be that in order to take the next step in the improvement of the environment, we must confront—much as we now do in education policy, welfare policy, or criminal justice policy—the fact of differentiated policy impacts. Not all people have benefited equally from past environmental regulation, nor have all shared the same burden of environmental risk. This inequality may never be completely resolved, just as inequalities in both educational opportunities and the application of criminal justice will probably never be eliminated. But specifics of any educational or criminal justice process have not only had to satisfy criteria of operational effectiveness, but also to confront (although seldom to resolve) the significance of the impact of these specific policies across diverse population groups. Until now, the environmental policy decision-making process has essentially ignored such impact issues. But times are now changing.

Of course, many could dispute the contention that the environmental-justice movement represents the beginning of a fundamental change in the nature of environmentalism in this country. There is a tendency to dismiss the movement as not a representation of an environmental policy paradigm shift but at best a moderately popular and probably short-lived community-based social movement. In a way, this doubting attitude could be compared with the attitude many had toward Medicare at its beginning. When Medicare first emerged, there were few, in or out of the health care professions, who recognized that it represented a fundamental change in the way health care would be provided in this country. Nor did these individuals or groups recognize what an immense impact Medicare would have on our economy and social organization. Medicare brought changes in the very way health care was provided in this country, who would provide it, and the relative roles of the major players in health care. Similarly, the introduction of environmental-justice concerns into the world of environmental policy will radically change the structure, attitudes, and major actors in that world. As Medicare does in the sphere of health care delivery, so too environmental justice could very well come to dominate, or significantly color, all environmental policy discussions. In the not-too-distant future, a time will come when the very thought of moving ahead with an environmental policy decision without first considering that policy's diverse impacts will be considered absurd.

Such changes are even more likely on the international environmental policy-making stage. As we move into the twenty-first century, it has become quite clear that many solutions to environmental problems cannot be unilaterally imposed. No one nation or group of nations alone can dictate international environmental policy outcomes. Yet international agreements or cooperative efforts often break down because one country or group of nations—usually from among the more developed, industrialized nations —seeks solutions that, while supposedly benefiting all, impose extraordinary economic, social, or health burdens on the less developed and powerless nations.

The solution to this international dilemma is not, as some have suggested, educating the developing nations on the importance of environmental protection. The solution may rather lie more in the developed nations being educated socially and economically about a different concept of environmental policy—a concept that recognizes differentiated policy impacts and respects the differences in countries' perspectives due to major disparities in income, education, and resources. Environmental justice at the international level will demand approaches and solutions that truly take into consideration the global dynamics of environmental activities.

Environmental justice is a virtually unknown environmental topic outside the United States. This is not due to lack of international importance. The international problems of environmental justice are just as significant, and in many cases more severe, than our domestic issues. It is rather due to many factors, including a less well structured international environmental agenda in general and the relatively slower development of grassroots community environmental groups in many countries, that environmental-justice issues are simply not yet significant international topics.

This relative lack of international standing suggests one possible key role the United States may adopt in the development of international environmental justice. What it would not and should not become is the world cop, imposing an American environmental vision on the rest of the globe. Such an approach is arrogant and, from a practical policy perspective, doomed to failure. Our more fitting role may be that of world educator and policy facilitator.

Providing environmental education, information, and opportunities for discussion is a role for which we are well qualified. No matter how bad environmental conditions may appear to be in this country, when it comes to environmental justice and the socioeconomic impact of environmental activities, the United States is usually well ahead of the rest of the world. Many of the environmental-justice lessons and experiences we have experienced and are now undergoing can serve as useful learning examples internationally. To most other governments, environmental justice is no more than a curiosity that has no standing as a serious policy issue. Yet environmental justice is not a topic whose importance can be ignored until every other environmental issue has been resolved. Failure to integrate environmental-justice issues into the current international environmental dialogue will guarantee political and economic troubles for future generations.

What's Important?

As I have mentioned several times throughout this book, it is both a strategic mistake and misguided policy analysis for both critics and supporters of environmental justice to focus all their energies on any particular case, whether local, national, or international. Too much time is spent on finding victims and villains; not enough time is spent on finding solutions. While recognizing the truth in the late Speaker of the House Tip O'Neil's statement that "all politics is local," it should not be forgotten that all social issues are global. Proving whether or not a company chose a particular location in a community or country on the basis of the color or income of the affected population does not change the larger equation. In that equation, all future environmental analysis, policy-making, and program design and implementation must incorporate social consequences into their agenda.

Besides the inevitability of the impact of environmental justice, the one thing to take away from this book is that we do not know enough about the many dimensions of environmental justice to more than fumble along, making the script up as we go. Yet none of the major players in this drama have the luxury of simply waiting until more understanding is achieved. If one followed to the letter, sparing no expense, all the suggestions and forms put forward in this book's chapters on measurement and analysis—or, for that matter, if one followed any other current analytic guide—for assessing environmental-justice conditions, the results would still be muddled. There is simply no current method, or collection of methods, for properly assessing the actual risk to a community of any particular environmental action or actions. Of particular difficulty is the joint and cumulative risk assessment that are needed in many environmental-justice situations. Nor is there promise of such an ability in the foreseeable future. Conversely, environmental decisions are currently being made, with or without good analytic assistance, that will have enormous differentiated impacts on different populations for a long time.

Affected populations and communities have legitimate needs for protection now, and they deserve far more consideration than environmental decision-makers have thus far given them. Accepting the imperfection of whatever solutions are suggested does not change the fact that a solution—a recommendation—is required. With flexibility and an awareness of the fallibility of the decision process, improvement in environmental justice can be realized, even under these conditions.

Appendix
Principles of Environmental Justice

WE, THE PEOPLE OF COLOR, gathered together at this multinational People of Color Environmental Leadership Summit to begin to build a national and international movement of people of color to fight the destruction and taking of our lands and communities, do hereby re-establish our spiritual interdependence to the sacredness of our Mother Earth; to respect and celebrate each of our cultures, languages and beliefs about the natural world and our roles in healing ourselves; to insure environmental justice; to promote economic alternatives which would contribute to the development of environmentally safe livelihoods; and to secure our political, economic and cultural liberation that has been denied for over 500 years of colonization and oppression, resulting in the poisoning of our communities and land and the genocide of our people, do affirm and adopt these Principles of Environmental Justice:

1. **Environmental justice** affirms the sacredness of Mother Earth, ecological unity and the interdependence of all species, and the right to be free from ecological destruction.

2. **Environmental justice** demands that public policy be based on mutual respect and justice for all peoples, free from any form of discrimination or bias.

3. **Environmental justice** mandates the right to ethical, balanced and responsible uses of land and renewable resources in the interest of a sustainable planet for humans and other living things.

4. **Environmental justice** calls for universal protection from nuclear testing, extraction, production and disposal of toxic/hazardous wastes

213

and poisons and nuclear testing that threaten the fundamental right to clean air, land, water and food.

5. **Environmental justice** affirms the fundamental right to political, economic, cultural, and environmental self-determination of all peoples.

6. **Environmental justice** demands the cessation of the production of all toxins, hazardous wastes, and radioactive materials, and that all past and current producers be held strictly accountable to the people for detoxification and containment at the point of production.

7. **Environmental justice** demands the right to participate as equal partners at every level of decision making including needs assessment, planning, implementation, enforcement, and evaluation.

8. **Environmental justice** affirms the right of all workers to a safe and healthy work environment, without being forced to choose between an unsafe livelihood and unemployment. It also affirms the right of those who work at home to be free from environmental hazards.

9. **Environmental justice** protects the rights of victims of environmental injustice to receive full compensation and reparations for damages as well as quality health care.

10. **Environmental justice** considers governmental acts of environmental injustice a violation of international law, the Universal Declaration on Human Rights, and the United Nations Convention on Genocide.

11. **Environmental justice** must recognize a special legal and natural relationship of Native Peoples to the U.S. government though treaties, agreements, compacts, and covenants affirming sovereignty and self-determination.

12. **Environmental justice** affirms the need for urban and rural ecological policies to clean up and rebuild our cities and rural areas in balance with nature, honoring the cultural integrity of all our communities, and providing fair access for all to the full range of resources.

13. **Environmental justice** calls for the strict enforcement of principles of informed consent, and a halt to the testing of experimental reproductive and medical procedures and vaccinations on people of color.

14. **Environmental justice** opposes the destructive operations of multinational corporations.

15. **Environmental justice** opposes military oppression, repression, exploitation of lands, people, and cultures, and other life forms.

16. **Environmental justice** calls for the education of present and future generations which emphasizes social and environmental issues, based

on our experiences and an appreciation of our diverse cultural perspectives.

17. **Environmental justice** requires that we, as individuals, make personal and consumer choices to consume as little of Mother Earth's
resources and to produce as little waste as possible; and to make the
conscious decision to challenge and re-prioritize our lifestyles to insure the health of the natural world for present and future generations.

Adopted, October 27, 1991
The First People of Color Environmental Leadership Summit
Washington, D.C.

Notes

Preface and Acknowledgments

1. Within the last few years, this situation has radically changed. A large and growing number of environmental policy, natural resource management, and public-policy schools, including my own, have begun formal offerings in this area. By the end of the 1990s, one was hard-pressed to find any such school that did not offer some consideration of environmental justice in its curriculum.
2. United Church of Christ Commission for Racial Justice and Public Data Access, *Toxic Wastes and Race in the United States* (New York: United Church of Christ Commission for Racial Justice, 1987).
3. Robert Bullard, "Solid Waste Sites and the Houston Black Community," *Sociological Inquiry* 53 (1983): 273–88.
4. This does not include a number of earlier works where the issues of distorted impact of environmental hazards or risks were observations in a larger work with a slightly different focus. For an excellent review of much of the existing literature until 1993 on differentiated-impact studies, see Benjamin A. Goldman's work for the National Wildlife Federation, *Not Just Prosperity* (prepared for the National Wildlife Federal Corporate Conservation Council, 1994).
5. Goldman, *Not Just Prosperity*.

1. Introduction

1. The information in the statements is taken from United Church of Christ Commission for Racial Justice and Public Data Access, *Toxic Wastes and Race*, and Marianne Lavelle and M. Coyle, "Unequal Protection: The Racial Divide on Environmental Law," *National Law Journal* 15 (September 21, 1992): 52–54.

   In the National Environmental Policy Act (NEPA) of 1969, the closest utterance of a concern for differentiated impact may be detected in the policies

217

and goals (section 101 [b] [2]), where the federal government is charged with the responsibility to "assure for all Americans safe, healthful, productive, and esthetically and culturally pleasing surroundings."

2. United Church of Christ Commission for Racial Justice and Public Data Access, *Toxic Wastes and Race.*

3. See, for example, Benjamin A. Goldman, *The Truth about Where You Live* (New York: Times Books/Random House, 1992).

4. Agency for Toxic Substances Disease Registry (ATSDR), *The Nature and Extent of Lead Poisoning in Children in the United States* (Atlanta, Ga.: Centers for Disease Control, 1988). Also see Olivia Carter-Pokras et al., "Blood Levels of Four- to Eleven-Year-Old Mexican American, Puerto Rican, and Cuban Children," *Public Health Reports* 105 (1990): 388–91.

5. In most states, formulas for the equalization of school funding are intended to counterbalance the weaker revenue base of poor-income districts. These formulas recognize that two districts of unequal wealth making the same "taxing effort" (i.e., applying the same rate of property tax or local income tax to themselves to support their schools) will generate vastly different total school revenues. The state attempts to level this playing field by allocating more money per student to the poor districts.

6. Ulrich Beck, *Risk Society,* trans. Mark Ritter (London: Sage Publications, 1992), p. 41; originally published as *Risikogesellschaft: Auf dem Weg in eine andere Moderne* (Frankfurt am Main: Suhrkamp, 1986).

2. Forms of Environmental Justice

1. National Council of Churches, *Policy Statement on Racial Justice* (New York: National Council of Churches, November 1984).

2. Cf. Arthur M. Okun, *Equality and Efficiency* (Washington, D.C.: Brookings Institution, 1975).

3. Another well-developed concept of equity is found in public finance applications. There, two major classes of equity forms, vertical equity and horizontal equity, serve as measurements of tax burden. For the first, the test is change in the tax burden across income classes. Tax schemes that produce increases, decreases, or no change in the percentage of income taxed are described as *progressive, regressive,* or *neutral,* respectively. Horizontal equity applies to changes in tax burden within an income category. For example, households with the same income, living in the same type of residential structure, within the same jurisdiction, and receiving the same government services should face the same tax burden. If they do not, horizontal inequity exists. Well-developed methodologies exist for the assessment of both vertical and horizontal equity; cf. Richard A. Musgrave and Peggy B. Musgrave, *Public Finance in Theory and Practice,* 3d ed. (New York: McGraw-Hill, 1980), and John L. Mikesell, *Fiscal Administration,* 5th ed. (Fort Worth, Tex.: Harcourt Brace College Publications, 1999).

4. Cf. J. LeGrand, *Equity and Choice* (London: Harper Collins, 1991).

5. U.S. Environmental Protection Agency (EPA), *Environmental Justice Report* (Washington, D.C.: Office of Environmental Equity, 1994).

6. U.S. EPA, *Final Guidance for Incorporating Environmental Justice Concerns in EPA's NEPA Compliance Analysis* (Washington, D.C.: U.S. EPA Office of Federal Activities, April 1998, section 1.1.1).

7. This, of course, does not deny that environmental justice can take on a different meaning and significance as the populations with standing are changed. That is, environmental justice takes on a much broader meaning if low-income populations are included under the environmental-justice umbrella, and yet broader if populations are given standing based just on geographic location, regardless of race, ethnicity, or income. Furthermore, the political acceptability of some environmental-justice initiatives admittedly does change with the changes in included populations. But political acceptability is not the same thing as conceptual or ideological inclusion.

8. Note the contrast between the conclusions of two reports: United Church of Christ Commission for Racial Justice and Public Data Access, *Toxic Wastes and Race*, and D. Anderton et al., "Hazardous Waste Facilities," *Evaluation Review* 18, no. 2 (1994): 123–40. A part of the difference in conclusion can be ascribed to different choices in geographic level of evaluation.

9. For study on the zip code level (in contrast to the study by Anderton et al., "Hazardous Waste Facilities," on the census tract level), see James Hamilton, "Testing for Environmental Racism: Prejudice, Profits, Political Power?" *Journal of Policy Analysis and Management* 14, no. 1 (1995): 107–32.

10. W. Bowen et al., "The Spatial Association between Race, Income, and Industrial Toxic Emissions in Cuyahoga County, Ohio" (paper prepared for the Annual Meeting of the Association of American Geographers, May 1993).

11. Principal among these other federal agencies with significant environmental responsibilities or impacts would be the U.S. Departments of Interior, Energy, Agriculture, Housing and Urban Development, and Defense.

12. Of course, there are environmental-justice problems for which these single-risk assessment methodologies are useful.

13. The command-and-control approach to environmental regulation involves specification of the type, method, and level of environmental remediation one must follow. Meet the command-and-control criteria, and the regulation is satisfied.

14. Vicki Been, "Market Forces, Not Racist Practices, May Affect the Siting of Locally Undesirable Land Uses," in *Environmental Justice*, ed. Jonathan S. Petrikin (San Diego: Greehaven Press, 1995), pp. 38–59.

15. U.S. EPA, Office of Pollution Prevention and Toxics, *Toxic Inventory and Emissions Reduction 1987–1990 in the Lower Mississippi Industrial Corridor Report* (Washington, DC: U.S. EPA, May 14, 1993).

16. Jim Motaualli, "Toxic Targets: Polluters that Dump on Communities of Color Finally Being Brought to Justice," *E Magazine* 9, no. 4 (July–August 1998): 28–41.

17. Louisiana Advisory Committee to the U.S. Commission on Civil Rights, *The Battle for Environmental Justice in Louisiana* (Kansas City: U.S. Commission on Civil Rights Regional Office, September 1993).

18. Keith Schneider, "Valley of Death for the Navajo Uranium Miner," *New York Times*, May 3, 1993, pp. A1, B10.

19. Karl Grossman, "Environmental Racism," *Crisis* 98, no. 4 (April 1991): 14–

21; and "From Toxic Racism to Environmental Justice," *E Magazine* 3 (May 1992): 28–35.

20. Margaret L. Knox, "Their Mother's Keeper," *Sierra* 78, no. 2 (March–April 1993): 51–57, 81–84.

21. United Church of Christ Commission for Racial Justice and Public Data Access, *Toxic Wastes and Race*.

22. Goldman, *Not Just Prosperity*.

23. United Church of Christ Commission for Racial Justice and Public Data Access, *Toxic Wastes and Race*.

24. Goldman, *Truth about Where You Live*.

25. Leslie A. Nieves, "Not in Whose Backyard? Minority Population Concentrations and Noxious Facilities Sites" (presentation for the American Academy of Science, Chicago, February 9, 1992), and "Regional Differences in the Potential Exposure of U.S. Minority Populations to Hazardous Facilities" (presentation for the Annual Meeting of the Regional Science Association, Chicago, November 19, 1992).

26. Results such as those of the Goldman and Nieves studies, with their high correlations between minority location and industrial hazards within counties, may also point to the greater concentration of minorities in urban areas, especially that of Asian Americans in cities on the two coasts.

27. For example, Goldman's *Truth about Where You Live* also shows that, excepting Hawaii, the ten U.S. counties with the highest percentage of Asian American inhabitants are all in California.

28. Lavelle and Coyle, "Unequal Protection."

29. It could be argued that location still plays a part in this type of problem as well, in that exposure to such hazards as lead poisoning is a direct result of living in older, poorly maintained housing.

30. P. West et al., "Minority Anglers and Toxic Fish Consumption: Evidence from a State-Wide Survey of Michigan," in *Race and the Incidence of Discourse*, ed. B. Bryant and Paul Mohai (Boulder, Colo.: Westview Press, 1992), pp. 108–22.

31. G. Friedman-Jimenez, "Achieving Environmental Justice: The Role of Occupational Health," *Fordham Urban Law Journal* 21, no. 3 (spring 1994): 605–31.

32. ATSDR, *Nature and Extent*.

33. P. Martin et al., "A Profile of California Farmworkers," *California Agriculture* 6 (1985): 16–18.

34. R. Mines et al., "Findings from the National Agricultural Workers Survey (NAWS) 1990, A Demographic and Employment Profile of Perishable Crop Workers" (Research Paper no. 1, Washington, D.C.: U.S. Department of Labor, Office of Program Economics, 1991).

35. M. Moses et al., "Environmental Equity and Pesticide Exposure," *Toxicology and Industrial Health* 9, no. 5 (1993): 913–59.

3. What Has Gone Before

1. Robert Gottlieb, *Forcing the Spring* (Washington, D.C.: Island Press, 1993), chap. 7; also Stephen Fox, *The American Conservation Movement* (Madison: University of Wisconsin Press, 1981).

2. This is particularly ironic, since writers such as Kent E. Portney have portrayed environmentalism, in its clash with economics and science, as a force of humanism. Humanism is a curious label to attach to a movement that appears to have completely overlooked the effects of its agenda on the socioeconomic conditions of the majority of humankind. See Portney, *Controversial Issues in Environmental Policy* (New York: Sage Publications, 1992).

3. Donald Snow, *Inside the Environmental Movement* (Washington, D.C.: Conservation Fund, Island Press, 1992), p. xxxiii.

4. Chapter 6 will more thoroughly discuss attitudes toward and treatment of people of color by federal environmental and natural resource agencies.

5. For a concise general survey of the different environmental philosophies and agendas, see David Pepper, *Modern Environmentalism* (London: Routledge, 1996), chap. 1.

6. As late as the early 1990s, many books on environmental policy had not recognized social justice, in virtually any of its manifestations, as a topic for debate among environmentalists.

7. Such uniform labeling does not ignore the influence of movements other than the preservation and conservation movement on the history and ethos of individual environmental organizations. For example, the deep ecologists clearly have been influenced by the radical antiwar movement of the late 1960s and early 1970s. Similarly, the antinuclear movement had a clear impact on some organizations. See, for example, Kirkpatrick Solo, *The Green Revolution* (New York: Hill and Wang, 1993).

8. Cf. N. Freudenberg and C. Steinsapir, "Not in Our Backyards: The Grassroots Environmental Movement," in *American Environmentalism*, ed. R. Dunlap and A. Mertig (Philadelphia: Taylor and Francis, 1992), pp. 27–37; Philip Shabecoff, "Environmental Groups Told They Are Racists in Hiring," *New York Times*, February 1, 1990, p. A20; and Charles Jordan and Donald Snow, "Diversification, Minorities and the Mainstream Environmental Movement," in *Voices from the Environmental Movement*, ed. Donald Snow (Washington, D.C.: Conservation Fund, Island Press, 1992), chap. 4.

9. Within the last five or six years, several of the larger national environmental organizations such as the Sierra Club have made major efforts to bring more people of color into their leadership circles.

10. See, for example, F. H. Buttle and W. L. Flinn, "Environmental Politics: The Structuring of Partisan and Ideological Cleavages in Mass Environmental Attitudes," *Sociological Quarterly* 17, no. 4 (1976): 477–90; also see Buttle and Flinn, "The Politics of Environmental Concern: The Impacts of Party Identification and Political Ideology on Environmental Concern," *Environment and Behavior* 10 (1978): 17–36, and L. W. Milbraith, *Environmentalists* (Albany: SUNY Press, 1984).

11. In an interesting survey study of the attitudes of groups as diverse as members of Earth First (a deep ecology group) and a group of sawmill workers where recent layoffs could be attributed to environmental activities, one of the largest areas of dissension between the two groups was in concern over environment protection versus social justice. Cf. W. Kempton et al., *Environmental Values in American Culture* (Cambridge, Mass.: MIT Press, 1995), chap. 8.

12. Murray Bookchin and Dave Foreman, *Defending the Earth*, ed. Murray Bookchin (Boston: South End Press, 1991), emphasis in original; introductory
comments by Steve Chase, p. 3.

13. One local chapter of a national environmental organization was frustrated
when it tried to engage youths from an inner-city community in local environmental activities. The environmental group could not understand why the
youths were not more enthusiastic about a tree-planting campaign for their
own neighborhoods (for which the environmental group was providing financing and expert guidance). In the end, the group abandoned the project when
several youths and their parents pointed out that in their neighborhood, the
trees would offer attractive hiding places for mugging and shooting ambushes.

14. Although minority concerns about the environment in general do not differ
much from the environmental concerns of others, minorities may place a
greater importance on specific policy items than is given to those items in the
agendas posited by mainstream environmental organizations.

15. Until the middle to late 1980s, environmental issues did not appear anywhere
in the annual reports or goals statements of major civil rights organizations
such as the Urban League and the NAACP.

16. Cf. Michael Cohen, *The History of the Sierra Club* (San Francisco: Sierra Club
Books, 1988). Of particular note is chap. 1, "Make the Mountains Glad,"
which chronicles the founding efforts of John Muir. For a direct source, see
John Muir, *Our National Parks* (Boston: Houghton Mifflin, 1901). An excellent source of the writings of the conservation movement's modern beginning
is R. Nash, "Part Two: The Progressive Conservation Crusade, 1901–1910,"
in *American Environmentalism: Readings in Conservation History*, 3d ed. (New
York: McGraw-Hill, 1990), pp. 69–71.

   Although he called himself one, John Muir was less a conservationist (as
the term is used today) than a preservationist, advocating not the "scientific"
and "wise use" of wilderness, which Gifford Pinchot championed, but rather
the maintenance of wilderness areas in a pristine, undisturbed condition.
Called a transcendental nature mystic, Muir had actually tried to appropriate
the term *conservationist*, which he imbued with a different meaning. However,
Pinchot, with his advocacy of government-based wise use, not only won the
major political battles of the time, but also gained ownership of the term *conservationist*. Cf. M. J. Killingsworth and J. S. Palmer, *Ecospeak* (Carbondale:
Southern Illinois University Press, 1992), and Fox, *American Conservation
Movement*.

17. J. M. Petulla, *American Environmental History*, 2d ed. (Columbus: Merrill,
1988).

18. Petulla, *American Environmental History*.

19. This is in sharp contrast to the late-medieval European attitude toward cities,
which is best summed up in the proverb, "Town air makes men free." For
people in that much earlier era, life in the countryside and wilderness was
viewed as a constant struggle for survival. People did not want to go back to
nature, but rather get as far away from it as possible. Cf. E. A. Johnson, *Some
Origins of the Modern Economic World* (New York: Macmillan, 1936).

20. Cf. David Pepper, *The Roots of Modern Environmentalism* (London: Croom Helm, 1984), pp. 86–87.
21. Cf. Milbraith, *Environmentalists.*
22. Pepper, *Roots of Modern Environmentalism.*
23. Pepper, *Roots of Modern Environmentalism.*
24. Cf. G. Rosen, *A History of Public Health* (Baltimore: John Hopkins University Press, 1993); expanded ed. of *A History of Public Health* (New York: MD Publications, 1958); J. Corn, *Environment and Health in Nineteenth Century America* (New York: Peter Lang, 1989).
25. Contrary to some impressions, urban gentrification and the return of the middle class to city life represents a minuscule percentage of annual middle-class housing relocation.
26. B. Devall, "Deep Ecology and Radical Environmentalism," in *American Environmentalism,* ed. R. Dunlap and A. Mertig (Philadelphia: Taylor and Francis, 1992), pp. 61–62.
27. Robert Bullard, *People of Color Environmental Groups Directory* (Riverside: University of California, Riverside, Department of Sociology, 1992). Updated online in 1996 and 2000. Available at: http://www.ejrc.cau.edu/poc2000.htm.

4. The Evolution of Environmental Justice as a Policy Issue

1. The most common agency response outside the EPA to the Presidential Executive Order of February 11, 1994, which established a requirement for the assessment of environmental justice for all federal agencies, was not to deny having an environmental-justice problem but rather to ask, What in the world is environmental justice?
2. The terms *public-policy issue* and *social issue* are used as a shorthand here for an issue that affects at least a major segment of society in a manner for which a public-sector response or solution is often pursued.
3. Robert Eyestone argues in *From Social Issues to Public Policy* (New York: Wiley, 1978) that "an issue arises when a public with a problem seeks or demands governmental action, and there is public disagreement over the best solution to the problem" (p. 3).
4. Group theory argues that public policy emerges from the struggle between competing interest groups. Individuals sharing some set of ideals or interests come together to increase their influence on public institutions. In group struggles, government is generally perceived as somewhat passive, acting more as a referee. Individuals are generally seen as relatively powerless, except as their interests are reflected in a particular group's interests and aims. Furthermore, because there is an inherent bias in interest group formulation toward the middle and upper classes, public policy naturally will reflect that same bias. Cf. Earl Latham, *The Group Basis of Politics* (New York: Octagon Books, 1965), and David Truman, *The Governmental Process* (New York: Knopf, 1951).
5. Elite theory starts from the premise that public policy is formed by the few for the many. From the perspective of elite theory, public policy reflects the

values not of collections of competing interest groups, but rather of the few who matter. Change will almost always be incremental rather than disjointed and sweeping, because the primary goal of change is as a measured response to challenges to the current system. Elite theory, much more than group theory, places considerably more emphasis on the importance of the individual, assuming that the individual is one of the elite. Cf. Thomas R. Dye and L. Harmon Zeigler, *The Irony of Democracy* (Belmont, Calif.: Wadsworth, 1970), and Robert Dahl, "Critique of the Ruling Elite Model," *American Political Science Review* 52 (June 1958): 463–69.

6. The approach of political systems to the policy process tries to be broad enough to accommodate all forces surrounding a problem. It is assumed that both individuals and groups generate demands or inputs into the political system, to which government then responds in the form of policy outputs. Cf. David Easton, "An Approach to the Analysis of Political Systems," *World Politics* 9 (April 1957): 383–400; *A Framework for Political Analysis* (Englewood Cliffs, N.J.: Prentice-Hall, 1965); and *A System Analysis of Political Life* (New York: Wiley, 1965).

7. Incrementalism describes how policy decision-makers arrive at particular outcomes. The incrementalists argue that policy-makers, in looking at a subset of all possible decisions, focus on those who make marginal changes in the existing structure. Incrementalism has a relatively reactive interpretation of policy-making, describing the typical policy-maker as having little desire to undertake radical, disjointed action without extraordinary motivation. A major voice in incrementalism is Charles Lindblom. See Lindblom, "The Science of 'Muddling Through,' " *Public Administration Review* 19 (1959): 79–88; and Edward J. Woodhouse, *The Policy-Making Process*, 3d ed. (Englewood Cliffs, N.J.: Prentice-Hall, 1993). See also Michael T. Hayes, *Incrementalism and Public Policy* (White Plains, N.Y.: Longman, 1992).

8. John Kingdon, among others, argues that policy-making is an evolutionary process in which only the fittest survive, in which the relative merits of an issue are subordinate to whether its time has come. Of particular importance is the presence of political entrepreneurs, who, when a window of opportunity opens, are able to lift a particular issue up from the garbage can of competing issues and utilize the window to bring about the desired change. Such windows open for a variety of reasons, such as technological breakthroughs, defining disasters, or electoral changes. Cf. John Kingdon, *Agendas, Alternatives, and Public Policies* (Boston: Little, Brown, 1984); "Agendas, Ideas, and Policy Change," in *New Perspectives on American Politics*, ed. Lawrence C. Dodd and Calvin Jillson (Washington, D.C.: CQ Press, 1994), pp. 215–29; and Michael Cohen et al., "A Garbage Can Model of Organizational Choice," *Administrative Science Quarterly* 17 (March 1972): 1–15.

9. The rational-analytic model of public policy-making, while recognizing that many of its steps cannot be completely realized, argues that the rationalizing steps it does take force decision-makers to delineate their true policy goals. At the same time, even when handicapped by inadequate data, the rational-comprehensive approach still provides decision-makers with a clearer under-

standing of the consequences of various solutions. Cf. David Weimer and Aidan R. Vining, *Policy Analysis,* 3d ed. (Englewood Cliffs, NJ: Prentice-Hall, 1999); Edith Stokey and Richard Zeckhauser, *A Primer for Policy Analysis* (New York: Norton, 1978); and Herbert Simon, "Rationality as Process and as Process of Thought," *Proceedings of the American Economic Association* 68 (May 1978): 1–16.

10. Cf. Amitai Etzioni, "Mixed Scanning: A 'Third' Approach to Decision-Making," *Public Administration Review* 27 (December 1967): 385–92.

11. Cf. Charles Jones, *An Introduction to the Study of Public Policy,* 3d ed. (Belmont, Calif.: Wadsworth, 1984); James Anderson, *Public Policy-Making,* 3d ed. (New York: Holt, Rinehart, and Winston, 1984); and Charles Bonser et al., *Policy Choices and Public Action* (Upper Saddle River, N.J.: Prentice-Hall, 1996), chap. 2.

12. Jones, *Introduction,* chap. 2.

13. For any number of reasons, an issue may languish in a particular stage of development, never moving forward or backward.

14. It is common for aspects of the same set of problems to occupy different stages of policy development.

15. Cf. Hayes, *Incrementalism and Public Policy*; Kingdon, *Agendas, Alternatives*; and Jones, *Introduction*.

16. Cf. Pepper, *Roots of Modern Environmentalism*.

17. The present description is an amalgamation of several descriptions of the policy issue process, including Grover Starling's life cycle of a social issue in *The Changing Environment of Business,* 3d ed. (Boston: PWS-Kent Publishing, 1988), p. 214; Jones, *Introduction;* and Bonser et al., *Policy Choices,* chap. 2.

18. Before this stage, one of the issues may have been viewed as part of some other larger matter. For example, location of a hazardous-waste facility in a minority neighborhood would have been viewed as merely part of a general urban-planning problem.

19. Cf. Lawrence Lynn, *Managing Public Policy* (Boston: Little, Brown, 1987), and several others, on nonequilibrium conditions.

20. Starling, *Changing Environment of Business*.

21. Anthony Downs, "Up and Down with Ecology: The 'Issue-Attention Cycle,'" *Public Interest* 28 (summer 1972): 38–50; quotation from pp. 39–40.

22. S. Cosgrove and A. Duff, "Environmentalism, Middle-Class Radicalism, and Politics," *Sociology Review* 28 (1980): 335–51; and W. Tucker, *Progress and Privilege: America in the Age of Environmentalism* (Garden City, N.Y.: Doubleday, 1982).

23. Barry Commoner, "Fundamental Causes of the Environmental Crisis," in *American Environmentalism,* ed. R. Nash, 3d ed. (New York: McGraw-Hill, 1990), pp. 206–14; quotation from p. 211; originally appeared in *Saturday Review* 53 (1970): 50–52, 62–64.

24. Fred Hoerger et al., "The Cumulative Impact of Health, Environmental, and Safety Concerns on the Chemical Industry During the Seventies," *Law and Contemporary Problems* 46, no. 3 (summer 1983): 59–107; quotation from p. 82.

25. Christopher Schroeder, "The Evolution of Federal Regulation of Toxic Sub-

stances," in *Government and Environmental Politics*, ed. Michael J. Lacey (London, Md.: Wilson Center Press, 1991).

26. Cf. John E. Ullman, ed., *The Suburban Economic Network* (New York: Praeger Publishers, 1977); of particular note are chap. 5, "Land Use," by Harvey Bloom; chap. 6, "Energy," by Jon P. V. Madden; and chap. 10, "Domestic Capital Equipment," by David Groelinger. Also see B. Wattenbert with R. Scammon, "The Suburban Boom," in *North American Suburbs*, ed. John Kramer (Berkeley: Glendessary Press, 1972), pp. 71–81.

27. Cf. World Bank, *The World Bank Atlas* (Washington, D.C.: International Bank for Reconstruction and Development/World Bank, 1992).

28. Aaron Wildousky, "No Risk Is the Highest Risk of All," *American Scientist* 67, no. 1 (January–February 1979): 32–37.

29. David Pepper, in *Roots of Modern Environmentalism*, argues that a major component of the late-nineteenth-century conservation and preservation movement was a repudiation of the urban environment in favor of country life.

30. In 1950, the median income in current dollars of a white family was $3,445 versus $1,869 for a black family, for a ratio of 54 percent, black to white. By the beginning of the modern environmental movement more than twenty years later, in 1972, the ratio had only improved to 59.4 percent, or $11,549 for a white family versus $6,864 for a black family. Harold Stanley and R. G. Niemi, *Vital Statistics on American Politics*, 5th ed. (Washington, D.C.: CQ Press, 1995), table 12-2, pp. 350–51.

31. By 1947, 33.2 percent of white men and 36.7 percent of white women twenty-five years and older had completed four years of high school or more, versus only 12.7 percent for black men and 14.5 percent for black women. At the beginning of the environmental movement in 1970, 57.2 percent of white men older than twenty-five years and 57.6 percent of white women had completed four years or more of high school, while blacks still lagged behind at 32.4 percent and 34.8 percent for black men and women, respectively. The black education attainment in 1970 was just about equal to that of whites in 1947, twenty-three years earlier. U.S. Bureau of the Census, *Current Population Reports*, "Educational Attainment in the United States . . ." (Washington D.C.: Government Printing Office, 1994), series P20, no. 476, pp. 96–98.

32. Rachel Carson, *Silent Spring* (Boston: Houghton Mifflin, 1962).

33. Benjamin Kleinberg, *Urban America in Transformation: Perspectives on Urban Policy and Development* (Thousand Oaks, Calif.: Sage Publications, 1995), chap. 5, "Restoring American Urbanism: Suburbanization and Urban Renewal." Also see Gary Tobin, "Suburbanization and the Development of Motor Transportation: Transportation Technology and the Suburbanization Process," in *The Changing Face of the Suburbs*, ed. B. Schwartz (Chicago: University of Chicago Press, 1976), pp. 95–111.

34. Although it could be argued that such a shift in risk and environmental burden may be justified if there were an accompanying payoff or compensating resource shift from the benefited group to the burdened group, none of the scant research in the area of environmental justice to date gives evidence of such a trade-off.

35. From 1947, just after the end of World War II, to 1960, minority median income increased in unadjusted dollars by 111 percent, while during the same period white families and individuals realized an 80 percent increase. From 1960 to 1970, minority median income increased 97 percent, versus an increase of 66 percent for whites. More significant, the ratio of minority income to white income changed from 48.2 percent in 1947 to 65.4 percent in 1970. Cf. U.S. Department of Commerce, Bureau of the Census, *Historical Statistics of the United States* (Washington, D.C.: Government Printing Office, 1975). After 1970, the progress of minority households is less clear. By 1989, the ratio of white income to that of African American households was 62.7 percent. Cf. U.S. Department of Commerce, Bureau of the Census, *1990 Census of Population*, CP-2-1 (Washington, D.C.: Government Printing Office, 1993).

36. These results come from, among other sources, a series of Roper Organization surveys conducted between the late 1980s and the present. The only minority group surveyed was the African American population.

37. For some discussion on misconceptions and the intransigent nature of the problem of poverty in this country, see William J. Wilson, *The Truly Disadvantaged* (Chicago: University of Chicago Press, 1987); of special note is chap. 6, "The Limited Visions of Race Relations and the War on Poverty."

38. Cf. National Academy of Public Administration (NAPA) Report to Congress, *Setting Priorities, Getting Results* (Washington, D.C.: NAPA, April 1995).

39. Upton Sinclair, *The Jungle* (1906; reprint, Urbana: University of Illinois Press, 1988).

40. Ralph Nader, *Unsafe at Any Speed* (New York: Grossman Publishers, 1965).

41. In 1900, 60.3 percent of the U.S. population lived in "rural territory"; within thirty years, the ratio had flipped and only 43.8 percent lived in rural areas. By 1970, the rural population of the country comprised only 26.5 percent of the total, and in 1990, the drop had continued, to 24.8 percent. See U.S. Department of Commerce, Bureau of the Census, *Historical Statistics*, dated 1975, and the U.S. Department of Commerce, Bureau of the Census, *U.S. Census of Population: Social and Economic Characteristics 1990* (PCO-1-B25; Washington, DC: Bureau of the Census, August 1992).

42. The 1906 Food and Drug Act's administration was assigned to what was then called the Bureau of Chemistry in the Department of Agriculture. The Bureau of Chemistry was reorganized in 1927 to separate the law enforcement component from agricultural research. It was formally named the Food and Drug Administration in 1931 and placed within what is now the Department of Health and Human Services. Cf. U.S. Food and Drug Administration, *The Story of the Laws behind the Labels*, part 1 (1906 Food and Drug Act, U.S. Food and Drug Administration, June 1981); available at: http://vm.cfsan.fda.gov/~/rd/history1.htm.

43. At the end of World War II, 28.2 million automobiles were registered in the U.S.; by 1970, that number had increased 216 percent to 89.3 million. During the same period, the population increased from 141.4 million to 204.9 million, or only 45 percent. U.S. Department of Commerce, Bureau of the Census, *Historical Statistics*.

44. There were only 8.1 million registered automobiles in the United States in 1920. Cf. U.S. Department of Commerce, Bureau of the Census, *Historical Statistics*.

45. This interpretation gives support to the elitist theory of policy formulation— that individuals, rather than the masses, have the key roles in policy creation.

46. U.S. General Accounting Office, *Siting of Hazardous Waste Landfills and Their Correlation with Racial and Economic Status of Surrounding Communities* (Washington, D.C.: General Accounting Office, 1983).

47. Bullard, "Solid Waste Sites."

48. See Joel A. Mintz, *Enforcement at the EPA* (Austin: University of Texas Press, 1995); of special note is chap. 7, "Lessons Learned: Congressional Oversight of EPA's Enforcement Approach in Broader Perspective"; see also Richard J. Lazarus, "The Tragedy of Distrust in the Implementation of Federal Environmental Law," and Richard J. Lazarus, "The Neglected Question of Congressional Oversight of EPA: Quis Custodiet Ipsos Custodes (Who Shall Watch the Watcher Themselves?)," in "Assessing the Environmental Protection Agency after Twenty Years: Law, Politics and Economics," ed. Christopher H. Schroeder and Richard J. Lazarus, a special issue of *Duke Journal of Law and Contemporary Problems* 54, no. 4 (fall 1991): 205–39.

49. P. Martin et al., "A Profile of California Farmworkers," *California Agriculture* 6 (1985): 16–18, and R. Mines et al., "Findings from the National Agricultural Workers Survey."

50. P. West et al., "Minority Anglers and Toxic Fish Consumption."

51. From Bullard's 1992 *People of Color Environmental Groups Directory;* of the fewer than 275 organizations and groups represented that were identified as having environmental-justice concerns, less than 30 percent were formed explicitly to deal with environmental problems.

52. In Bullard's 1994 *People of Color Environmental Groups Directory*, the number of organizations or groups with environmental justice as one objective had increased to over four hundred, and the percentage that were formed with an environmental primary mission had increased to 38 percent.

53. Roger Cobb and Charles Elder, *Participation in American Politics* (Boston: Allyn and Bacon, 1972), p. 14.

54. As of 1999, one of the few, if only, major academic research institutes devoted exclusively to the study of environmental justice is the one at Clark University in Atlanta, directed by Robert Bullard.

55. Cf. Kevin Trenberth, ed., *Climate System Modeling* (Cambridge: Cambridge University Press, 1992).

56. Hayes, *Incrementation and Public Policy,* chap. 4, "The Unequal Struggles"; Jones, *Introduction,* chap. 4, "Getting Problems to Government."

57. Cf. James E. Anderson, *Public Policy-Making*; Hayes, *Incrementation and Public Policy,* chap. 8, "The Policy Process: Some Generalizations."

58. For a more developed treatment of the power and policy choices, see R. Heineman et al., *The World of the Policy Analyst* (Chatham, N.J.: Chatham House, 1990).

59. For two views, see E. Savas, *Privatizing the Public Sector* (Chatham, N.J.: Chatham House, 1982); C. Lehman, "The Revolution of Saints: The Ideology

of Privatization and Its Consequences for Public Lands," in *Selling the Federal Forests,* ed. Adrian E. Gamache (Seattle: University of Washington, 1984).

60. Cf. Heineman et al., *World of the Policy Analyst,* chap. 5, "Policy Analysis and the Political Arena."

61. The roots of the Program Planning and Budgeting System (PPBS) actually go back several decades before the 1960s. However, it was the system's success, particularly at the Department of Defense under Robert McNamara, that prompted President Johnson to mandate its use in all federal agencies. Cf. Thomas Lynch, *Public Budgeting in America,* 2d ed. (Englewood Cliffs, N.J.: Prentice-Hall, 1985), chap. 2.

62. See E. Harper et al., "Implementation and Use of PPB in Sixteen Federal Agencies," in *Perspectives on Budgeting,* ed. Allen Schick, 2d ed. (Washington, D.C.: American Society for Public Administration, 1987), pp. 90–100.

63. In the first Nixon administration, in the early 1970s, the mandate for general federal usage of PPBS was lifted. Cf. Lynch, *Public Budgeting in America.*

64. After Executive Order 12898 of 1994, probably the most significant federal environmental-justice regulatory or legislative actions were the EPA's issue in April 1995 of "The EPA's Environmental Justice Strategy" in response to the mandates of Executive Order 12898, followed by EPA's *Interim Guidance* in March 1997 and *Final Guidance* in April 1998, and finally EPA's controversial *Interim Guidance for Investigating Title VI Administrative Complaints Challenging Permits,* issued March 1998. This last regulatory initiative focuses on investigating possible violations of Title VI of the Civil Rights Act of 1964 in the issuance of pollution control permits. Violations complaints would be processed by the recently created (1998) Office of Civil Rights. Title VI violation in the permitting process is a relatively narrow area of environmental-justice concern. Furthermore, a Title VI civil rights violation is a rather difficult matter to prove, and EPA has to date not met with much success. However, the very threat of Title VI action appears to have generated a great deal of attention and interest in environmental-justice issues on the part of many permit holders or applicants.

5. Misconceptions about Minority Attitudes toward Environmental Issues

1. U.S. Office of Personnel Management, *Federal Civilian Workforce Statistics* (Washington, D.C.: U.S. Office of Personnel Management, September 30, 1992). For a more complete discussion of the representation of minority professionals in federal environmental and natural resource protection and management agencies, see Chapter 6.

2. U.S. EPA, *Headquarters Cultural Diversity Survey: Final Report* (Washington, D.C.: U.S. EPA, May 1993).

3. On a scale of 1 (strongly disagree) to 4 (strongly agree), this statement had a mean value of 3.35. Snow, *Inside the Environmental Movement,* pp. 78–79.

4. Among the volunteer leaders, the mean value for this statement was 3.21 (where 4 was "strongly agree"). The survey asked respondents to supply information about the groups (e.g., salary ranges, educational backgrounds, funding

sources). Curiously, no attempt was made to ascertain the racial composition of the membership of the staff. Nor did a companion survey of academic programs for the management of environmental and natural resources include any queries about the composition of the student body or the faculty. Snow, *Inside the Environmental Movement*, pp. 126–27.

5. M. R. Hersey and P. B. Hill, "Is Pollution a White Thing? Racial Differences in Pre-Adult Attitudes," *Public Opinion Quarterly* 41, no. 4 (1977–78): 439–58.

6. Milbraith, *Environmentalists*, p. 78.

7. One clear range of differences exists between the national environmental organizations with a focus on issues such as global warming and air quality and local grassroots groups with a sharper focus on the siting of hazardous-waste facilities and threats of toxic pollution.

8. Cf. P. Mohai, "Black Environmentalism," *Social Science Quarterly* 71, no. 4 (December 1990): 744–65; Judi A. Caron, "Environmental Perspectives of Blacks: Acceptance of the New Environmental Paradigm," *Journal of Environmental Education* 20, no. 3 (spring 1989): 21–26; Francis O. Adeola, "Environmental Hazards, Health, and Racial Inequity in Hazardous Waste Distribution," *Environment and Behavior* 26, no. 1 (January 1994): 99–126.

9. A more recent survey of Illinois communities actually notes that environmental issues rank lower than crime, drugs, employment, and economic welfare for both majority and minority populations. Cf. Jaap Voss, "The Role of Local Planners and Decision-Makers in the Occurrence of Environmental Justice" (Ph.D. dissertation, Florida Atlantic University, 1997).

10. The more profound results from several Roper surveys were that although interest in environmental issues was high for both blacks and whites, neither group had a good understanding or awareness of the specifics of environmental issues. Nor were environmental causes and organizations something to which the American population as a whole was willing to devote large amounts of time or money.

11. Roper Organization, *Roper Report 90-10* (fieldwork October 1990, report January 1991).

12. Daniel Krause, "Environmental Consciousness: An Empirical Study," *Environment and Behavior* 25, no. 1 (January 1993): 126–42; quotation from p. 134.

13. Mohai, "Black Environmentalism," 755.

14. Roper Organization, *Roper Report 91-9* (fieldwork September 1991, report January 1992).

15. Victor Fisher et al., *A Survey of the Public's Attitude toward Soil, Water, and Renewable Resource Conservation Policy* (Washington, D.C.: U.S. Government Printing Office, 1980).

16. Caron, "Environmental Perspectives of Blacks"; Riley Dunlap and K. Van Liere, "The New Environment Paradigm," *Journal of Environmental Education* 9, no. 4 (summer 1978): 10–19.

17. Adeola, "Environmental Hazards," p. 123.

18. Roper Organization, *Roper Report 93-2* (fieldwork January 1993, report April 1993).

19. Roper Organization, *Roper Report 92-4* (fieldwork March 1992, report July 1992).

20. This, of course, may be an artifact of the power of the two nuclear choices and the other very undesirable choice, a state prison, simply crowding out all other choices from this relatively small set.

21. Roper Organization, *Roper Report 92-10* (fieldwork October 1992, report March 1993).

22. The Jaap Voss study of Illinois communities also found African American concerns over a number of community-based environmental issues to be slightly higher than those of the white populations. Cf. Voss, "Role of Local Planners and Decision-Makers."

23. Roper Organization, *Roper Report 83-8* (fieldwork August 1983, report October 1983).

24. Shabecoff, "Environmental Groups Told They Are Racists in Hiring."

25. Robert Norman, Director of Human Resources, National Audubon Society, quoted in Shabecoff, "Environmental Groups," p. 23.

26. Gottlieb, *Forcing the Spring,* chaps. 4 and 5; and R. Dunlap and A. Mertig, eds., *American Environmentalism* (Philadelphia: Taylor and Francis, 1992).

27. Gottlieb, *Forcing the Spring;* Snow, *Inside the Environmental Movement.*

28. Robert Bullard and B. Wright, "The Politics of Pollution: Implications for the Black Community," *Phylo* 47 (1986): 71–78.

29. Snow, *Inside the Environmental Movement.*

30. There was an inclusion of some environmental impact concerns after the mid-1990s.

31. Roper Organization, *Roper Report 89-4* (fieldwork March 1989, report June 1989).

32. Roper Organization, *Roper Report 89-4.*

6. The EPA

1. Senior executive service (SES) is the civil service career-grading system that replaced the GS supergrades (such as GS-16, GS-17, and GS-18) in most federal agencies in the 1970s.

2. President Bill Clinton issued his Environmental Justice Executive Order in February 1994, requesting, among other actions, an assessment of environmental justice from all federal agencies, with the EPA serving as the coordinating agency. One of the most common responses to the order from non-EPA federal agencies who called the EPA's Office of Environmental Justice was not the typical complaint about not having enough time to do a good job. The most common response from these agencies was, Just what is environmental justice?

3. U.S. EPA, *Summary of the Meeting of the National Environmental Justice Advisory Council: Washington, D.C., July 25-26, 1995* (Washington, D.C.: U.S. EPA, 1995), p. ES-1.

4. One indication that this commitment is perhaps window dressing are the comments attributed to EPA staffers that input from the National Environmental Justice Advisory Council (NEJAC) was neither as useful nor as extensive as EPA had expected. To this, a NEJAC member countered that EPA staffers simply did not want to listen to what NEJAC was saying, so they looked else-

where for a story they liked better. See *Inside EPA* (Arlington, Va.: Inside Washington Publishers, November 4, 1994).

5. U.S. General Accounting Office, *Siting of Hazardous Waste Landfills* (Washington, D.C.: Government Printing Office, 1983).

6. By the end of 1994, none of these agencies had made more than a token effort at addressing environmental-justice issues within their sphere of operations.

7. This charge discounts recent Office of Research and Development assertions that, ex post facto, describes numerous research projects as having an environmental-justice component.

8. Although the present discussion focuses on federal agencies, the problem is at least as serious at the state and local levels.

9. Jack Lewis, "The Birth of EPA," *EPA Journal* 12, no. 9 (November 1985): 6–11.

10. U.S. EPA, *U.S. Environmental Protection Agency Oral History Interview—1: William D. Ruckelshaus* (Washington, D.C.: U.S. EPA, January 1993); U.S. EPA, *U.S. Environmental Protection Agency Oral History Interview—2: Russell E. Train* (U.S. EPA, July 1993); Lewis, "Birth of EPA"; Robert Cahn and Patricia Cahn, "The Environmental Movement since 1970," *EPA Journal* 12, no. 9 (November 1985): 31–35.

11. William D. Ruckelshaus, quoted in "Views from the Former Administrators," *EPA Journal* 12, no. 9 (November 1985): 12.

12. Comments of Russell E. Train, quoted in "Views from the Former Administrators," p. 12.

13. U.S. EPA, *Headquarters Cultural Diversity Survey*.

14. See C. Frankfort-Nachimas and D. Nachimas, *Research Methods in the Social Sciences*, 4th ed. (New York: St. Martin's Press, 1992).

15. Frankfort-Nachimas and Nachimas, *Research Methods*.

16. Statistics in this chapter are taken from U.S. Environmental Protection Agency, *Headquarters Cultural Diversity Survey: Final Report* (Washington, D.C.: U.S. EPA, May 1993); U.S. Office of Personnel Management, Office of Workforce Information Central Personnel Data File Report as of September 30, 1982, O PM, 1982; U.S. Office of Personnel Management, Office of Workforce Information Central Personnel Data File Report as of September 30, 1988, O PM, 1988; and U.S. Office of Personnel Management, Office of Workforce Information Central Personnel Data File Report as of September 30, 1992, O PM, 1992.

17. African Americans (61 percent), Hispanics (54 percent), Asian or Pacific Islanders (57 percent), and Native Americans (58 percent) disagreed.

18. U.S. Office of Personnel Management, *Federal Civilian Workforce Statistics*.

19. In the actual survey, the three items did not appear in this order.

20. U.S. EPA, *Headquarters Cultural Diversity Survey*.

21. As of 1994, no other federal agency has published any other survey information on diversity of attitudes like that conducted by the EPA.

22. Only statistics for the overall federal executive agencies, the EPA, and a combined environmental and natural resource agencies classification are provided.

23. A large part of the EPA is organized into units representing the various pos-

sible media of pollution, such as water, air, and soil. Historically, this form of organization was intended only as a temporary device until a transition to a more integrated organizational form could be developed. Unfortunately, once established, the media form proved very difficult to alter.

24. The members of the office, from the director down to the newest intern, are dedicated to refining and promulgating the principles of environmental justice. They are the vanguard of the environmental-justice movement within the EPA. Not too many years ago, many of these individuals were the voices ignored by the environmental mainstream.

Part II: Policy Analysis of Environmental Justice

1. By *corrective* I mean government regulation to change or modify a specific set of circumstances in which actions are regulated with accompanying sanctions for noncompliance. This is in contrast to government endorsement of a particular behavior such as supporting the principle of environmental justice via its proclamation as an issue of human dignity, without any accompanying specifics.

2. For those who perceive this argument as a subtle attempt at championing a market-based solution to most environmental problems, they are incorrect. It is not subtle.

3. The National Academy of Public Administration provides some useful insights on the problem of a purely local solution, citing the prefederal intervention strategy of "smokestack chasing," which was a common local economic development approach that led to a "race to the bottom." See NAPA's Report to Congress, *Setting Priorities*.

4. There is a question whether "purely" social policy problems, with no associated market factor, exist for public-policy issues.

7. Environmental Justice through the Lens of Policy Analysis

1. In other words, the impacts, often today unknown, of various environmental policies and activities, across socioeconomic classifications, are uniform in neither type nor magnitude.

2. From David Weimer and Aidan R. Vining, *Policy Analysis,* 2d ed. (Englewood Cliffs, N.J.: Prentice-Hall, 1992), chap. 3.

3. This does not, of course, deny the fact that much of actual public policy reflects reaction to any number of political and social forces—of which failure of private solutions is just one. Public administration literature, some of which was reviewed in earlier chapters, provides many other reasons for government involvement. What is implied is that the principle of government intervention as a final response, rather than as a first response, should be a working premise.

4. By *policy actions,* I refer to specific policy initiatives that result in some collection of activity changes or creations.

5. Aside from the current impossibility of actually assessing environmental burden, insisting on the equal distribution of an environmental risk implies that everyone assigns exactly the same priority to avoiding exposure to that risk. The most elementary economics text would point out that, given nonidentical preferences and wealth, maintaining an equal distribution of burden means some individuals will face greater risk than they would prefer, and some less. Thus, a policy that insists on an equal distribution of burden actually leads to a less than optimal outcome for society as a whole.

6. This is called Pareto efficiency, named after the Italian economist and sociologist Vilfredo Pareto (1848–1923).

7. In 1975, Okun summarized this trade-off between efficiency and other societal goals quite well in *Equality and Efficiency.*

8. One of the classic causes of this is the existence of externalities in a market transaction. That is, some of the positive or negative consequences of a transaction are not borne by the parties to the exchange. There is spillover of transaction effects to parties outside the transaction, including the general community.

9. Unequal distribution of effects simply means differences in effects across populations. Inequity of distribution implies that according to some criteria of justice, the impact (social, economic, health) of these effects is not fair.

10. Community members are paying for increased health problems resulting from actions traceable to the firm's activities. Thus, part of the cost of firm production is in increased health costs to community members, which the firm is not paying.

11. Charles L. Schultze, *The Public Use of Private Interest* (Washington, D.C.: Brookings Institution, 1977), p. 32.

12. As the uncertainty associated with space exploration has decreased, as policy analysts predicted, more and more "routine" space ventures in the 1980s and 1990s have been undertaken by private firms. Likewise, the level of government intervention in environmental activities can almost certainly decrease as the uncertainty is reduced.

13. U.S. EPA, *Brownfields Action Agenda* (Washington D.C.: U.S. EPA, January 25, 1995).

14. Cf. U.S. Congress, Office of Technology Assessment, *The State of the States on Brownfields* (Washington, D.C.: U.S. Office of Technology Assessment, June 1995).

15. Cf. N. Leigh, "Focus: Environmental Constraints to Brownfield Redevelopment," *Economic Development Quarterly,* no. 4 (November 1994): 323–28; and K. Yount and P. Meyer, "Bankers, Developers, and New Investment in Brownfields Sites: Environmental Concerns and the Social Psychology of Risk," *Economic Development Quarterly* 8, no. 4: 338–44.

16. A public good is one that has the characteristics of nonexclusion and joint use. With a public good such as national defense, individual members of the community cannot be excluded from enjoying it, and an individual's use of it does not diminish anyone else's use.

17. The right to clean air or water has only recently evolved as a property right.

Further, assignment of the "clean" right to communities is itself an even newer development.

18. There are those, of course, who would argue that some fundamental rights do not change, regardless of circumstances or political currency.

19. In C. Jung et al., "The Coarse Theorem in a Rent-Seeking Society," *International Review of Law and Economics* 15, no. 3 (1995): 259–68, a strong argument is made that the assignment of rights is where the real bargaining will often occur and that the solution of many of the problems associated with efficient outcomes depends more on this process than on any subsequent activities.

20. When the transaction costs are large and asymmetric, however, the prospect of an efficient outcome decreases.

8. The Measurement of Environmental Justice

1. Even though market failure or distributional inequity may be the principal reasons for government intervention in these cases, other reasons for intervention, such as correcting a social or moral injustice, may still apply.

2. For example, both United Church of Christ Commission for Racial Justice and Public Data Access, *Toxic Wastes and Race*, and an earlier study by R. Bullard ("Solid Waste Sites") used zip codes as the base unit for geographic measurement. Subsequently, a number of writers have argued that smaller units, such as census tracts or block groups, are far more appropriate for local measurement than are zip codes or, worse, county-level units. See Vicki Been, "Locally Undesirable Land Uses in Minority Neighborhoods: Disproportionate Siting or Market Dynamics," *Yale Law Journal* 103, no. 6 (April 1994): 1383–422; and Anderton et al., "Hazardous Waste Facilities."

3. Cf. Weimer and Vining, *Policy Analysis*, chap. 8, "Landing on Your Feet: How to Confront Policy Problems."

4. This assumption, of course, ignores the environmental-justice problem of asymmetric information. In other words, these residents may have entered the community with a weak understanding of the environmental risks they would face.

5. There of course is a question about the validity of EPA's use of separate media assessment. Among others, NAPA in their report to Congress in April 1995 (*Setting Priorities*) raised questions about this approach.

6. This approach avoids the danger of drawing a conclusion that is more an artifact of the methodology than a reliable assessment of the situation. For an extreme example of such a case, see A. Charnes et al., "A Goal Programming/Constrained Regression Review of the Bell System Breakup," *Management Science* 34, no. 1 (1988): 1–26.

7. The division of environmental-justice methods into six categories is arbitrary. Others might separate individual methodologies, such as decision analysis and risk assessment, into their own individual classes, or interpret univariate and multivariate statistics as two separate methodological types. The six divisions simply highlight the major differences and utilities of various methodologies

and the different measurement and methodological needs that various classes of environmental-justice problems demand.

9. A New Way of Looking at the Same Old Numbers

1. In areas such as the environmental-justice issue of communities feeling that environmentally risky ventures are their only viable economic development alternative—for example, Native American tribal councils considering the permitting of nuclear waste storage on their lands—quantitative evaluation methods play only a tiny role, if any, in investigation of the issue. In the same manner, the environmental-justice issue of unusual pesticide exposure on predominantly Latino farm workers involves exploring the justice of specific practices, not whether a specific analytic methodology can better describe events.

2. See, for example, United Church of Christ Commission for Racial Justice and Public Data Access, *Toxic Wastes and Race*; and James Hamilton, "Politics and Social Costs: Estimating the Impact of Collective Action on Hazardous Waste Facilities," *Rand Journal of Economics* 24, no. 1 (spring 1993): 101–25. An excellent summary and listing of sixty-four empirical studies on environmental justice performed and published since the late 1960s is Goldman's *Not Just Prosperity*.

3. In a cross-sectional comparison study, data are collected on a number of communities at a single point in time.

4. As noted earlier, the single-risk-factor approach can provide useful insights if one assumes a high degree of correlation among types of environmental hazards. Actually, a high correlation among certain types of environmental hazards may not be that unusual. If one assumes that much the same economic, political, and technical factors operate in determining the location of many of these risk factors, then it should not be surprising if such high correlations do exist. Naturally, when there is a high degree of correlation among the environmental risk factors, then merely evaluating the distribution characteristics of a single factor does provide a rapid assessment of the total environmental-justice conditions.

5. See, for example, United Church of Christ Commission for Racial Justice and Public Data Access, *Toxic Wastes and Race*; and Bullard, "Solid Waste Sites."

6. Employing data envelopment analysis (DEA) does not appreciably improve the problem of the geographic unit of measurement.

7. United Church of Christ Commission for Racial Justice and Public Data Access, *Toxic Wastes and Race*.

8. John Hird, "Environmental Policy and Equity: The Case of Superfund," *Journal of Policy Analysis and Management* 12, no. 2 (1993): 323–43.

9. Anderton et al., "Hazardous Waste Facilities."

10. Bowen et al., "Spatial Association."

11. Been, "Locally Undesirable Land Uses."

12. Bullard, "Solid Waste Sites."

13. The original two works were A. Charnes et al., "Measuring the Efficiency of

Decision-Making Units," *European Journal of Operational Research* 2, no. 6 (1978): 429–44; and A. Charnes et al., "Evaluating Program and Managerial Efficiency: An Application of Data Envelopment Analysis to Program Follow Through," *Management Science* 27, no. 6 (1981): 668–87.

14. More recent development in DEA theory and application has allowed many and assorted separations of performance efficiency effects into categories such as managerial versus program efficiency and technical production versus cost efficiency. Exploration of these many and diverse variations of DEA within the context of environmental justice or equity assessment, although useful, I save for others.

15. Charnes et al., "Efficiency of Decision-Making Units."

16. T. Ahn et al., "Some Statistical and DEA Evaluations of Relative Efficiencies of Public and Private Institutions of Higher Learning," *Socio-Economic Planning Sciences* 22, no. 6 (1988): 259–69.

17. Two examples of the many articles written by these two authors, either as a pair or with others, are A. Bessent and W. Bessent, "Determining the Comparative Efficiency of Schools through Data Envelopment Analysis," *Education Administration Quarterly* 16, no. 2 (1980): 57–75; and A. Bessent et al., "An Application of Mathematical Programming to Assess Productivity in the Houston Independent School District," *Management Science* 29, no. 12 (1982): 1355–67.

18. E. L. Rhodes and L. Southwick, "Variations in Public and Private University Efficiency," *Applications of Management Science* 7 (1993): 145–70.

19. H. D. Sherman, "Hospital Efficiency and Evaluation: Empirical Test of a New Technique," *Medical Care* 22, no. 10 (October 1984): 922–38.

20. R. D. Banker et al., "Some Models for Estimating Technical and Scale Inefficiencies in Data Envelopment Analysis," *Management Science* 30, no. 9 (1984): 1078–92.

21. A. Y. Lewin et al., "Evaluating the Administrative Efficiency of Courts," *Omega* 10, no. 4 (1982): 401–11.

22. E. L. Rhodes, "An Exploratory Analysis of Variations in Performance among U.S. National Parks," in *Measuring Efficiency*, ed. R. H. Silkman, New Directions for Program Evaluation, no. 32 (San Francisco: Jossey Bass, 1986), pp. 47–71.

23. R. Fare et al., "The Relative Efficiency of Illinois Electric Utilities," *Resources and Energy* 5, no. 4 (1983): 349–67.

24. D. C. Adolpson et al., "Railroad Property Valuation Using Data Envelopment Analysis," *Interfaces* 19, no. 3 (1989): 18–26.

25. W. F. Bowlin, "An Intertemporal Assessment of the Efficiency of Air Force Accounting and Finance Offices," *Research in Government and Nonprofit Accounting* 5 (1989): 293–310.

26. In an interesting example of parallel development, the alternative statistical procedures were largely developed during the same time period as DEA (the late 1970s and early 1980s). Some have argued that this largely independent development simply represents a response of quantitative policy analysts and researchers to an increasing demand for methods to assess government activities.

27. Some of the other multiple-outcome evaluation approaches make such high data and information demands that outside of some very special cases, these approaches are almost impossible to actually apply to real-world public-sector problems. For a comparison of these procedures, see A. Charnes et al., "Comparison of DEA and Existing Ratio and Regression Systems for Effecting Efficiency Evaluation of Regulated Electric Cooperatives in Texas," *Research in Government and Nonprofit Accounting* 5 (1989): 187–210; and C. L. Lovell and P. Schmidt, "A Comparison of Alternative Approaches to the Measurement of Productive Efficiency," in *Applications of Modern Production Theory*, ed. A. Dogramaci and R. Fare (Boston: Kluwer Academic Publishers, 1987), pp. 3–32.

28. For an excellent primer that walks the reader through the more standard DEA procedures, see R. D. Banker et al., "An Introduction to Data Envelopment Analysis with Some of Its Models and Their Uses," *Research on Government and Nonprofit Accounting* 5 (1989): 125–63.

29. This small computational convention is needed because the DEA procedure assumes that more of what is defined as output is better than less, so the environmental conditions were restated as environmental quality conditions in which more is preferred to less, versus environmental risk conditions, for which less is preferred to more.

30. There are special circumstances where the maximum DEA value can be greater than one, but such special manipulations need not concern the present work.

31. In the environmental-justice case, various community characteristics such as mean household income and level of educational attainment are factor inputs into the DEA calculations, and the vector of resultant environmental quality factors based on a conversion of a set of environmental risk measurements are the DEA-evaluated outcomes. That is, the environmental quality condition measurements will be conversions into positive terms of a collection of negatively oriented environmental risk factors. For example, the original environmental risk factors could be the number of Toxic Release Inventory (TRI) sites within a given geographic range and the number of National Priority List (NPL) sites from the larger Superfund site list within a certain range of a neighborhood.

Next choose the maximum number in each category of TRI sites and NPL sites observed for any community area in the observation set. Now, subtract from the maximum number, the actual number of TRI and NPL sites observed in each community. In this way, a community with no TRI and no NPL sites would have an environmental quality measurement for two factors that were maximally high. In contrast, those communities with sites would have their environmental quality number reduced from the average.

32. An alternative DEA approach to the environmental quality assessment question is to incorporate all influencing factors, including race and income, in the initial DEA calculations. The resultant DEA values could then be divided at some designated threshold into those that have relatively high minority populations and those that do not. A statistical comparison could then be made between the two groups to ascertain whether differences in DEA values are

significant or not. One major disadvantage of such a single-stage approach is determining the dividing line threshold on the basis of community racial composition. Whatever the criteria, it will be somewhat arbitrary. In contrast, the proposed two-staged approach makes no predeterminations about racial neighborhood membership. This latter approach allows variable racial composition along some continuum.

33. The value 1.0 has been added to the quality measurement in order to satisfy a DEA algorithm requirement that does not permit the entry of a value of 0 for any factor.

34. The *t* statistic is a measurement of the likelihood that the coefficient value is other than 0. A *t* statistic of −2.479 says that at least 95 percent of the time, it is correct to assume that minority percentage has an effect on the DEA value.

Part III: A Case, a Summary, and Some Conclusions

1. See K. Viscusi, "Reforming OSHA Regulation of Workplace Risks," in *Regulatory Reform*, ed. L. Weiss and M. Klass (Boston: Little, Brown, 1986), pp. 234–68.

10. A Case of Environmental Justice

1. Emelle, Alabama, is the site of the country's largest hazardous-waste processing facility.

2. The 1980 population of 13,212 dropped to 12,604 in 1990, a decline of 4.6 percent, which followed a loss of 7.5 percent between 1970 and 1980. U.S. Department of Commerce, Bureau of the Census, *U.S. Census Population: General Population Characteristics 1990*, CP-1-26 (Bureau of the Census, May 1992), and *U.S. Census of Population: General Population Characteristics 1980*, PCO-1-B25 (Bureau of the Census, August 1992).

3. *U.S. Census of Population: General Population Characteristics 1980*, PCO-1-B25.

4. In the 1990s, when Mississippi had an unemployment rate of about 8 percent, Noxubee County's unemployment ranged between 14 and 15 percent. For the county's African Americans during the same period, the unemployment rate was more than 22.4 percent. U.S. Department of Commerce, Bureau of the Census, *U.S. Census of Population: Social and Economic Characteristics 1990*.

5. *Indiana University Foundation Annual Report, 1997–1998* (Bloomington: Indiana University Press, 1999), p. 2.

6. The foundation also received a management fee for each fund's investments. In 1992, these management fees totaled $893,265. In 1992, the foundation owned a total of $14,967,475 in real estate assets, of which $8,307,567 was from the endowment and similar funds category. Cf. *Indiana University Foundation Annual Report*.

7. Colin Crawford, *Uproar at Dancing Rabbit Creek* (Reading, Mass.: Addison-Wesley, 1996), p. 121.

8. Although at least two other potential buyers had made tentative overtures

during the previous two years, they lacked the financial means to back their interests.

9. There is no evidence that the racial composition of Noxubee county—or even the concept of environmental racism—ever came up at board discussions when the Federated Technologies, Inc. (FTI), proposal was considered. Nor did there appear to be any evidence that the board anticipated any alarm on campus over their actions, and accordingly, no special attempt was made to disseminate information on this decision.

10. Owen Williams, "Records: Federated is Toxic Waste Rookie," *Meridian (Miss.) Star*, October 23, 1990, pp. 1A, 11A.

11. Open letter from FTI to Noxubee County residents, November 6, 1990.

12. Letter to Indiana University Foundation Board from Hughes Environmental Systems–Federated Technologies of Mississippi, Inc., December 17, 1992.

13. John Coffey, "Hazardous Waste Firm Picks Brooksville as Site," *Commercial Dispatch* (Columbus, Miss.), October 16, 1990, p. 1A.

14. John Coffey, "Firm's Waste Disposal Experience Unclear," *Commercial Dispatch*, October 24, 1990, p. 1A.

15. During subsequent negotiations, FTI formed a subsidiary, Golden Triangle Resources, to meet the solid-waste management responsibilities of their agreement (mainly for collection and hauling) via a regional solid-waste authority with municipal participation. Stated reasons by FTI for this were to allow Noxubee County to participate "in an economical fashion with other counties in the region for the collective handling of solid waste," and also because FTI had been informed by the State Department of Environmental Quality that under state law the solid-waste management obligation may be the responsibility of a separate entity. Letter, Edward Netherland to the law offices of Maxey, Pigott, Wann, and Begley, November 25, 1992.

16. Vicky Oswat, "Hazardous Waste Approval Expected," *Commercial Dispatch*, March 24, 1991, pp. 1A, 10A.

17. Press release, Federated Technologies of Mississippi, Inc., August 16, 1991.

18. "Hughes and FTI Ratify Noxubee Partnership," *Commercial Dispatch*, August 18, 1991, p. 1A.

19. EPA Region IV comprises Kentucky, Tennessee, North and South Carolina, Alabama, Georgia, Florida, and Mississippi.

20. The Comprehensive Environmental Response, Compensation, and Liability Act as amended 1986, section 104 (c) (9).

21. Quotation by Linda Thomas in *Columbus (Miss.) Republic*, December 2, 1990.

22. Nick Schneider, "Mississippi Woman Fights IU Land Sale," *Herald-Times* (Bloomington, Ind.), April 22, 1991, pp. A3, A7.

23. The full text of the Student Environmental Action Coalition resolution appears in the appendix to this chapter.

24. Press release from student groups, Indiana University, Bloomington, Ind., August 8, 1992.

25. Student press release, August 5, 1992.

26. Teri Klassen, "Many in Noxubee Support Land Sale," *Herald-Times*, July 26, 1992, p. C1.

27. Klassen, "Many in Noxubee," p. C3.
28. "Bankhead Rejects Racist Argument," *Commercial Dispatch*, July 28, 1992, p. 9A. Bankhead's statements were publicized widely in several newspapers in both Mississippi and Indiana.
29. Chris Rickett, "IUSA Accuses Noxubee Bidder of Bribery to Gain County Favor," *Indiana Daily Student* (Bloomington, Ind.), August 6, 1992, p. 10.
30. See the Appendix for the full text of the October 9 resolution.
31. The students submitted a 130-page document to the Indiana University Foundation for reply by FTI in October. The FTI response several months later was even larger: a three-ring binder containing five hundred pages of clippings, testimonials, and one-sentence replies.
32. Tracy Huber, "IUF Board Members to Stand by Land Sale," *Indiana Daily Student*, October 26, 1992, p. 1.
33. Teri Klassen, "Fighting Land Sale Costly," *Herald-Times*, November 5, 1992, p. C1.
34. Ashraf Kholil, "Council Drawn into Land Sale Feud," *Indiana Daily Student*, January 21, 1993, p. 1.
35. Letter to Indiana University Foundation from Representative William A. Crawford, January 28, 1993.
36. Teri Klassen, "Company May Yet Buy Site for Incinerator," *Herald-Times*, February 9, 1993.
37. Steven Higgs, "No Sale in Noxubee," *Herald-Times*, April 2, 1993, pp. A1, A9.
38. Crawford, *Uproar at Dancing Rabbit Creek*, p. 312.
39. Crawford, *Uproar at Dancing Rabbit Creek*, p. 329.

11. Policy Directions and Recommendations

1. Cf. Jordan and Snow, "Diversification, Minorities."
2. The definition of population or work environment also extends to non-area-specific environmental-justice conditions, such as farm workers' exposure to pesticides, or the consumption of contaminated fish.
3. After the fall elections of 1994, which put a Republican cadre in control of both the House and Senate, this legislative process ground to an abrupt halt.
4. Cf. L. Weiss and M. Klass, eds., *Regulatory Reform: What Actually Happened* (Boston: Little, Brown, 1986).
5. Such intergenerational equity problems are a major feature of the management of public-sector capital debt, and it may be wise to consider employing some of the devices developed in that field to compensation and cost distribution in the area of environmental justice. Cf. Mikesell, *Fiscal Administration*.
6. A conference in the fall of 1993 at the EPA on the application of geographic information systems (GIS) to the environmental-justice problem provided some interesting examples of the type of information provision that does not serve this need. By using the powerful but also complex and cumbersome ARCINFO system, several of the regional participants demonstrated mapping and information tools that they proposed making available to the public. Unfortunately, the ARCINFO system simply does not lend itself to either in-

quiry or manipulation by anyone not well versed in the intricacies of its codes and commands. Maps that take, even with assistance, twenty minutes to an hour to generate, and data retrieval exercises that presuppose a familiarity with a wealth of GIS jargon are accessible to no one but a small handful of well-trained analysts. That the tool was actually forwarded as a community access device is not an indication of, in this case, the EPA's opposition to citizen involvement, but rather symptomatic of the EPA's inexperience in ensuring such involvement. Although not subsequently adapted, the fact that this was even suggested highlights the lack of appreciation of the provision of community-level information.

7. The use of multiple analytic procedures and multiple scales avoids the danger of conclusions that are based more on the level of analysis than on the process being studied.

8. And it will exist for most acts. Any act or regulation, if subjected to enough comparative analysis across enough different dimensions, will show uneven application or outcome along some dimension. Such unevenness does not automatically provide proof of bias or discrimination.

9. A neutral factor of influence would be a group characteristic that is not only beyond the control of the agency, but also does not represent any preexisting social bias. For example, more Superfund cleanups in minority communities may be solved through containment rather than through the more traditional avenue of removal. This could represent bias or problems of inequity. However, let us assume that the EPA, which has primary responsibility for Superfund administration, determines that (1) the disproportionate outcomes arise because hazardous materials sites in minority communities generally belong to a class of sites for which disposal containment is the preferred response; and furthermore that (2) the disproportionate distribution of this type of sites resulted from historical patterns, not bias in site classification by the EPA. In examining equity consistency, however, the EPA should go one step further and (3) determine whether or not recommendation for disposal as the preferred solution for this type of site was itself influenced by its location or who lived near it.

10. For example, a number of environmental and health regulations have a provision allowing citizen access to information on specific sites, outcomes, or conditions. Unfortunately, the manner of either obtaining that information or using the specific resultant data assumes that the citizen already possesses skills or convenience of location that many community members simply do not possess.

11. U.S. EPA, *Final Guidance*.

12. For example, the guidance notes that holding public meetings in unfamiliar surroundings such as government buildings and luxury hotels does not encourage community participation. It also notes the problem of work schedule conflicts, which may impact working-class communities more than others. Most important, it is noted that technically complex issues need to be presented in a manner understandable to the intended community participants.

13. U.S. EPA, *Interim Guidance for Investigating Title VI Administrative Complaints Challenging Permit* (Washington, D.C.: U.S. EPA, March 1998).

14. "Low income" is not a class covered under this guidance; U.S. EPA, *Interim Guidance for Investigating Title VI Administrative Complaints Challenging Permit*, pp. 8–10.

15. U.S. EPA, *EPA Brownfields Assessment Demonstration Pilots* (Washington, D.C.: U.S. EPA, October 1996). By spring 1998, there were 227 such pilot projects.

16. U.S. EPA, *Brownfields Title VI Case Studies: Summary Report* (Washington, D.C.: U.S. EPA, June 1999).

17. The 1989 United Nations–sponsored *Basel Convention on the Control of Transboundary Movement of Hazardous Wastes and Their Disposal (1989)* (Washington, D.C.: Government Printing Office, 1991) directly addressed issues of protecting countries from receiving unwanted shipments of wastes. Furthermore, its 1995 amendment banned export of hazardous waste from developed countries to less developed countries. Although the United States was a prime mover behind the original Basel Conventions, by the beginning of the new century, it has yet to ratify either the original convention or its more restrictive export ban, the 1995 amendment. Cf. Mary Tiemann, "Waste Trade and the Basel Convention: Background and Update" (report for Congress, Congressional Research Service, December 1998).

18. Zada Lipman, "Trade in Hazardous Waste: Environmental Justice versus Economic Growth" (presented at the Environmental Justice Global Ethics Fourth 21st Century Conference, Melbourne, Australia, October 1–3, 1997; available at: http://www.arbld.unimelb.edu.au/enjust/papers/allpapers/lipman/home.htm). A. Viv, "Toxic Trade with Africa," in *Environmental Science Technology Journal* 23 (1989), also includes comment on hazardous-waste trade with Haiti by Organization for Economic Cooperation and Development countries. For additional notes, see "Dumping by Another Name in Haiti" (Friends of the Earth, 1999; available at: http://www.foe.co.uk/foei/tes/link8.html).

19. Cf. Bullard, "Solid Waste Sites"; Been, "Locally Undesirable Land Uses."

20. M. Lavelle, "Community Profile, Chicago: An Industrial Legacy," *National Tax Journal*, special issue on environmental racism, p. 53; Lavelle and Coyle, "Unequal Protection," pp. 52, 54; M. Ervin, "The Toxic Doughnut," *Progressive* 56 (January 1992): 15.

21. Cf. R. Bullard, "Ecological Inequalities and the New South: Black Communities under Siege," *Journal of Ethnic Studies* 17, no. 4 (winter 1990): 101–15, and *Dumping in Dixie* (Boulder, Colo.: Westview Press, 1990).

22. From its inception, this local prerogative focused on the issue of exclusion and problems of externalities. Cf. J. Heilbrun, "Site Rent, Land-Use Patterns, and the Form of the City," chap. 6 of *Urban Economics and Public Policy*, 2d ed. (New York: St. Martin's Press, 1981); and D. Judd, *The Politics of American Cities*, 2d ed. (Boston: Little, Brown, 1984), chap. 6.

23. Wayland Gardner, *Government Finance* (Englewood Cliffs, N.J.: Prentice-Hall, 1978).

24. Cf. M. Suksi, *Bringing in the People* (Dordrecht: Martinus Nijhoff, 1993), chaps. 1 and 2; and J. Allswang, *California Initiatives and Referendums 1912–1990* (Los Angeles: Edmund G. "Pat" Brown Institute of Public Affairs, 1991).

25. Hard copy is also available.

12. Environmental Justice

1. Until the executive order on environmental justice in 1994, many of these "other" federal agencies did not even see themselves as having environmental policy responsibilities. If nothing else, environmental justice has helped expand and redefine what an environmental policy issue is and who must become involved.

Bibliography

Adeola, Francis O. "Environmental Hazards, Health, and Racial Inequity in Hazardous Waste Distribution." *Environment and Behavior* 26, no. 1 (January 1994): 99–126.

Adolpson, D. C., G. C. Cornia, and L. C. Walters. "Railroad Property Valuation Using Data Envelopment Analysis." *Interfaces* 19, no. 3 (1989): 18–26.

Agency for Toxic Substances Disease Registry (ATSDR). *The Nature and Extent of Lead Poisoning in Children in the United States: A Report to Congress.* Atlanta, Ga.: Centers for Disease Control, 1988.

Ahn, T., A. Charnes, and W. W. Cooper. "Some Statistical and DEA Evaluations of Relative Efficiencies of Public and Private Institutions of Higher Learning." *Socio-Economic Planning Sciences* 22, no. 6 (1988): 259–69.

Allswang, J. *California Initiatives and Referendums 1912–1990: A Survey Guide to Research.* Los Angeles: Edmund G. "Pat" Brown Institute of Public Affairs, 1991.

Anderson, James E. *Public Policy-Making.* 3d ed. New York: Holt, Rinehart, and Winston, 1984.

Anderton, D., et al. "Hazardous Waste Facilities: 'Environmental Equity' Issues in Metropolitan Area." *Evaluation Review* 18, no. 2 (1994): 123–40.

Banker, R. D., A. Charnes, and W. W. Cooper. "Some Models for Estimating Technical and Scale Inefficiencies in Data Envelopment Analysis." *Management Science* 30, no. 9 (1984): 1078–92.

Banker, R. D., A. Charnes, W. W. Cooper, J. Swartz, and D. Thomas. "An Introduction to Data Envelopment Analysis with Some of Its Models and Their Uses." *Research on Government and Nonprofit Accounting* 5 (1989): 125–63.

Beck, Ulrich. *Risk Society: Toward a New Modernity.* Translated by Mark Ritter. London: Sage Publications, 1992. Originally published as *Risikogesellschaft: Auf dem Weg in eine andere Moderne.* Frankfurt am Main: Suhrkamp, 1986.

Been, Vicki. "Locally Undesirable Land Uses in Minority Neighborhoods: Disproportionate Siting or Market Dynamics." *Yale Law Journal* 103, no. 6 (April 1994): 1383–422.

——. "Market Forces, Not Racist Practices, May Affect the Siting of Locally Undesirable Land Uses." In *Environmental Justice*, ed. Jonathan S. Petrikin, pp. 38–59. San Diego: Greehaven Press, 1995.

Bessent, A., and W. Bessent. "Determining the Comparative Efficiency of Schools through Data Envelopment Analysis." *Education Administration Quarterly* 16, no. 2 (1980): 57–75.

Bessent, A., W. Bessent, J. Kennington, and B. Regan. "An Application of Mathematical Programming to Assess Productivity in the Houston Independent School District." *Management Science* 29, no. 12 (1982): 1355–67.

Bonser, Charles, E. B. McGregor Jr., and Clinton V. Oster Jr. *Policy Choices and Public Action*. Upper Saddle River, N.J.: Prentice-Hall, 1996.

Bookchin, Murray, and Dave Foreman. *Defending the Earth: A Dialogue between Murray Bookchin and Dave Foreman*. Edited by Murray Bookchin. Boston: South End Press, 1991.

Bowen, W., et al. "The Spatial Association between Race, Income, and Industrial Toxic Emissions in Cuyahoga County, Ohio." Paper prepared for the Annual Meetings of the Association of American Geographers, May 1993.

Bowlin, W. F. "An Intertemporal Assessment of the Efficiency of Air Force Accounting and Finance Offices." *Research in Government and Nonprofit Accounting* 5 (1989): 293–310.

Bullard, Robert. "Ecological Inequalities and the New South: Black Communities under Siege." *Journal of Ethnic Studies*, 17, no. 4 (winter 1990): 101–15.

——. *Dumping in Dixie: Race, Class, and Environmental Quality*. Boulder, Colo.: Westview Press, 1990.

——. *People of Color Environmental Groups Directory*. Riverside: University of California, Riverside, Department of Sociology, 1992. Updated online in 1996 and 2000. Available at: http://www.ejrc.cau.edu/poc2000.htm.

——. "Solid Waste Sites and the Houston Black Community." *Sociological Inquiry* 53 (1983): 273–88.

Bullard, Robert, and B. Wright. "The Politics of Pollution: Implications for the Black Community." *Phylon* 47 (1986): 71–78.

Buttle, F. H., and W. L. Flinn. "Environmental Politics: The Structuring of Partisan and Ideological Cleavages in Mass Environmental Attitudes." *Sociological Quarterly* 17, no. 4 (1976): 477–490.

——. "The Politics of Environmental Concern: The Impacts of Party Identification and Political Ideology on Environmental Concern." *Environment and Behavior* 10 (1978): 17–36.

Cahn, Robert, and Patricia Cahn. "The Environmental Movement since 1970." *EPA Journal* 12, no. 9 (November 1985): 31–35.

Caron, Judi A. "Environmental Perspectives of Blacks: Acceptance of the New Environmental Paradigm." *Journal of Environmental Education* 20, no. 3 (spring 1989): 21–26.

Carson, Rachel. *Silent Spring*. Boston: Houghton Mifflin, 1962.

Carter-Pokras, Olivia, et al. "Blood Levels of Four- to Eleven-Year-Old Mexican American, Puerto Rican, and Cuban Children." *Public Health Reports* 105 (1990): 388–91.

Charnes, A., W. W. Cooper, D. Divin, T. W. Ruefli, and D. Thomas. "Comparison of DEA and Existing Ratio and Regression Systems for Effecting Efficiency Evaluation of Regulated Electric Cooperatives in Texas." *Research in Government and Nonprofit Accounting* 5 (1989): 187–210.

Charnes, A., W. W. Cooper, and E. L. Rhodes. "Measuring the Efficiency of Decision-Making Units." *European Journal of Operational Research* 2, no. 6 (1978): 429–44.

———. "Evaluating Program and Managerial Efficiency: An Application of Data Envelopment Analysis to Program Follow Through." *Management Science* 27, no. 6 (1981): 668–87.

Charnes, A., W. W. Cooper, and T. Sueyoshi. "A Goal Programming/Constrained Regression Review of the Bell System Breakup." *Management Science* 34, no. 1 (1988): 1–26.

Cobb, Roger, and Charles Elder. *Participation in American Politics: The Dynamics of Agenda Building.* Boston: Allyn and Bacon, 1972.

Cohen, Michael. *The History of the Sierra Club: 1892–1970.* San Francisco: Sierra Club Books, 1988.

Cohen, Michael, James March, and Joham Olsen. "A Garbage Can Model of Organizational Choice." *Administrative Science Quarterly* 17 (March 1972): 1–15.

Commoner, Barry. "Fundamental Causes of the Environmental Crisis." In *American Environmentalism: Readings in Conservation History*, ed. R. Nash. 3d ed., pp. 206–14. New York: McGraw-Hill, 1990. Originally appeared in *Saturday Review* 53 (1970): 50–52, 62–64.

Comprehensive Environmental Response, Compensation, and Liability Act, as amended 1986. Section 104 (c) (9).

Corn, J. *Environment and Health in Nineteenth Century America: Two Case Studies.* New York: Peter Lang, 1989.

Cosgrove, S., and A. Duff. "Environmentalism, Middle-Class Radicalism, and Politics." *Sociology Review* 28 (1980): 335–51.

Crawford, Colin. *Uproar at Dancing Rabbit Creek.* Reading, Mass.: Addison-Wesley, 1996.

Dahl, Robert. "Critique of the Ruling Elite Model." *American Political Science Review* 52 (June 1958): 463–69.

Devall, B. "Deep Ecology and Radical Environmentalism." In *American Environmentalism: The U.S. Environment Movement, 1970–1990*, ed. R. Dunlap and A. Mertig, pp. 51–62. Philadelphia: Taylor and Francis, 1992.

Downs, Anthony. "Up and Down with Ecology—The 'Issue-Attention Cycle.' " *Public Interest* 28 (summer 1972): 38–50.

"Dumping by Another Name in Haiti." Friends of the Earth, 1999. Available at: http://www.foe.co.uk/foei/tes/link8.html.

Dunlap, Riley, and K. Van Liere. "The New Environment Paradigm." *Journal of Environmental Education* 9, no. 4 (summer 1978): 10–19.

Dunlap, R., and A. Mertig, eds. *American Environmentalism: The U.S. Environmental Movement, 1970–1990.* Philadelphia: Taylor and Francis, 1992.

Dye, Thomas R., and L. Harmon Zeigler. *The Irony of Democracy.* Belmont, Calif.: Wadsworth, 1970.

Easton, David. "An Approach to the Analysis of Political Systems." *World Politics* 9 (April 1957): 383–400.

——. *A Framework for Political Analysis.* Englewood Cliffs, N.J.: Prentice-Hall, 1965.

——. *A System Analysis of Political Life.* New York: Wiley, 1965.

Ervin, M. "The Toxic Doughnut." *Progressive* 56 (January 1992): 15.

Etzioni, Amitai. "Mixed Scanning: A 'Third' Approach to Decision-Making." *Public Administration Review* 27 (December 1967): 385–92.

Eyestone, Robert. *From Social Issues to Public Policy.* New York: Wiley, 1978.

Fare, R., S. Grosskopf, and J. Logan. "The Relative Efficiency of Illinois Electric Utilities." *Resources and Energy* 5, no. 4 (1983): 349–67.

Fisher, Victor, John Boyle, Mark Schulman, and Michael Bucuvala. *A Survey of the Public's Attitude toward Soil, Water, and Renewable Resource Conservation Policy.* Washington, D.C.: Government Printing Office, 1980.

Fox, Stephen. *The American Conservation Movement: John Muir and His Legacy.* Madison: University of Wisconsin Press, 1981.

Frankfort-Nachimas, C., and D. Nachimas. *Research Methods in the Social Sciences.* 4th ed. New York: St. Martin's Press, 1992.

Freudenberg, N., and C. Steinsapir. "Not in Our Backyards: The Grassroots Environmental Movement." In *American Environmentalism: The U.S. Environmental Movement, 1970–1990,* ed. R. Dunlap and A. Mertig, pp. 27–37. Philadelphia: Taylor and Francis, 1992.

Friedman-Jimenez, G. "Achieving Environmental Justice: The Role of Occupational Health." *Fordham Urban Law Journal* 21, no. 3 (spring 1994): 605–31.

Gardner, Wayland. *Government Finance.* Englewood Cliffs, N.J.: Prentice-Hall, 1978.

Goldman, Benjamin. *Not Just Prosperity: Achieving Sustainability with Environmental Justice.* Prepared for the National Wildlife Federal Corporate Conservation Council, 1994.

——. *The Truth about Where You Live: An Atlas for Action on Toxins and Mortality.* New York: Times Books/Random House, 1992.

Gottlieb, Robert. *Forcing the Spring: The Transformation of the American Environmental Movement.* Washington, D.C.: Island Press, 1993.

Grossman, Karl. "Environmental Racism." *Crisis* 98, no. 4 (April 1991): 14–21.

Hamilton, James. "Politics and Social Costs: Estimating the Impact of Collective Action on Hazardous Waste Facilities." *Rand Journal of Economics* 24, no. 1 (spring 1993): 101–25.

——. "Testing for Environmental Racism: Prejudice, Profits, Political Power?" *Journal of Policy Analysis and Management* 14, no. 1 (1995): 107–32.

Harper, E., F. Kramer, and A. Rouse. "Implementation and Use of PPB in Sixteen Federal Agencies." In *Perspectives on Budgeting,* ed. Allen Schick, pp. 90–100. 2d ed. Washington, D.C.: American Society for Public Administration, 1987.

Hayes, Michael T. *Incrementalism and Public Policy.* White Plains, N.Y.: Longman, 1992.

Heilbrun, J. *Urban Economics and Public Policy.* 2d ed. New York: St. Martin's Press, 1981.

Heineman, R., W. Bluhm, S. Peterson, and E. Kearny. *The World of the Policy Analyst: Rationality, Values, and Politics.* Chatham, N.J.: Chatham House, 1990.

Hersey, M. R., and P. B. Hill. "Is Pollution a White Thing? Racial Differences in Pre-Adult Attitudes." *Public Opinion Quarterly* 41, no. 4 (1977–1978): 439–58.

Hird, John. "Environmental Policy and Equity: The Case of Superfund." *Journal of Policy Analysis and Management* 12, no. 2 (1993): 323–43.

Hoerger, Fred, William H. Beamer, and James S. Hansen. "The Cumulative Impact of Health, Environmental, and Safety Concerns on the Chemical Industry During the Seventies." *Law and Contemporary Problems* 46, no. 3 (summer 1983): 59–107.

Indiana University Foundation Annual Report, 1997–1998. Bloomington: Indiana University, 1999.

Inside EPA. Arlington, Va.: Inside Washington Publishers, November 4, 1994.

Johnson, E. A. *Some Origins of the Modern Economic World.* New York: Macmillan, 1936.

Jones, Charles. *An Introduction to the Study of Public Policy.* 3d ed. Belmont, Calif.: Wadsworth, 1984.

Jordan, Charles, and Donald Snow. "Diversification, Minorities and the Mainstream Environmental Movement." In *Voices from the Environmental Movement: Perspectives for a New Era,* ed. Donald Snow, pp. 71–109. Washington, D.C.: Conservation Fund, Island Press, 1992.

Judd, D. *The Politics of American Cities.* 2d ed. Boston: Little, Brown, 1984.

Jung, C., K. Krutilla, W. K. Viscusi, and R. G. Boyd. "The Coarse Theorem in a Rent-Seeking Society." *International Review of Law and Economics* 15, no. 3 (1995): 259–68.

Kempton, W., J. S. Bosterm, and J. Hartley. *Environmental Values in American Culture.* Cambridge, Mass.: MIT Press, 1995.

Killingsworth, M. J., and J. S. Palmer. *Ecospeak: Rhetoric and Environmental Politics in America.* Carbondale: Southern Illinois University Press, 1992.

Kingdon, John. *Agendas, Alternatives, and Public Policies.* Boston: Little, Brown, 1984.

———. "Agendas, Ideas, and Policy Change." In *New Perspectives on American Politics,* ed. Lawrence C. Dodd and Calvin Jillson, pp. 215–29. Washington, D.C.: CQ Press, 1994.

Kleinberg, Benjamin. *Urban America in Transformation: Perspectives on Urban Policy and Development.* Thousand Oaks, Calif.: Sage Publications, 1995.

Knox, Margaret L. "Their Mother's Keeper." *Sierra* 78, no. 2 (March–April 1993): 51–57, 81–84.

Krause, Daniel. "Environmental Consciousness: An Empirical Study." *Environment and Behavior* 25, no. 1 (January 1993): 126–42.

Latham, Earl. *The Group Basis of Politics.* New York: Octagon Books, 1965.

Lavelle, M. "Community Profile, Chicago: An Industrial Legacy." *National Tax Journal* 15, no. 3 (September 21, 1992), p. 53. Special issue on environmental racism.

Lavelle, Marianne, and M. Coyle. "Unequal Protection: The Racial Divide on Environmental Law." *National Law Journal* 15 (September 21, 1992): S2, S4.

Lazarus, Richard J. "The Neglected Question of Congressional Oversight of EPA: Quis Custodiet Ipsos Custodes (Who Shall Watch the Watcher Themselves?)." In *Assessing the Environmental Protection Agency after Twenty Years: Law, Politics and Economics*, ed. Christopher H. Schroeder and Richard J. Lazarus. Special issue of *Duke Journal of Law and Contemporary Problems* 54, no. 4 (fall 1991): 205–39.

———. "The Tragedy of Distrust in the Implementation of Federal Environmental Law." In *Assessing the Environmental Protection Agency after Twenty Years: Law, Politics and Economics*, ed. Christopher H. Schroeder and Richard J. Lazarus. *Duke Journal of Law and Contemporary Problems* 54, no. 4 (fall 1991): 311–74.

LeGrand, J. *Equity and Choice: An Essay in Economics and Applied Philosophy.* London: HarperCollins, 1991.

Lehman, C. "The Revolution of Saints: The Ideology of Privatization and Its Consequences for Public Lands." In *Selling the Federal Forests*, ed. Adrian E. Gamache. Seattle: University of Washington, 1984.

Leigh, N. "Focus: Environmental Constraints to Brownfield Redevelopment." *Economic Development Quarterly*, no. 4 (November 1994): 323–28.

Lewin, A. Y., R. C. Morey, and T. J. Cook. "Evaluating the Administrative Efficiency of Courts." *Omega* 10, no. 4 (1982): 401–11.

Lewis, Jack. "The Birth of EPA." *EPA Journal* 12, no. 9 (November 1985): 6–11.

Lindblom, Charles. "The Science of 'Muddling Through.' " *Public Administration Review* 19 (1959): 79–88.

Lipman, Zada. "Trade in Hazardous Waste: Environmental Justice versus Economic Growth." Presented at the Environmental Justice Global Ethics Fourth 21st Century Conference, Melbourne, Australia, October 1–3, 1997. Available at: www.arbld.unimelb.edu.au/enjust/papers/allpapers/lipman/home.htm.

Louisiana Advisory Committee to the U.S. Commission on Civil Rights. *The Battle for Environmental Justice in Louisiana . . . Government, Industry, and the People.* Kansas City: U.S. Commission on Civil Rights Regional Office, September 1993.

Lovell, C. L., and P. Schmidt. "A Comparison of Alternative Approaches to the Measurement of Productive Efficiency." In *Applications of Modern Production Theory: Efficiency and Productivity*, ed. A. Dogramaci and R. Fare, pp. 3–32. Boston: Kluwer Academic Publishers, 1987.

Lynch, Thomas. *Public Budgeting in America.* 2d ed. Englewood Cliffs, N.J.: Prentice-Hall, 1985.

Lynn, Lawrence. *Managing Public Policy.* Boston: Little, Brown, 1987.

Martin, P., R. Mines, and A. Diaz. "A Profile of California Farmworkers." *California Agriculture* 6 (1985): 16–18.

Mikesell, John. *Fiscal Administration: Analysis and Application for the Public Sector.* 5th ed. Fort Worth, Tex.: Harcourt Brace College Publications, 1999.

Milbraith, L. *Environmentalists: Vanguard for a New Society.* Albany: SUNY Press, 1984.

Mines, R., S. Gabbard, and B. Boccalandro. "Findings from the National Agricultural Workers Survey (NAWS) 1990, a Demographic and Employment Profile of Perishable Crop Workers." Research Paper no. 1. Washington, D.C.: U.S. Department of Labor, Office of Program Economics, 1991.

Mintz, Joel A. *Enforcement at the EPA: High Stakes and Hard Choices.* Austin: University of Texas Press, 1995.

Mohai, P. "Black Environmentalism." *Social Science Quarterly* 71, no. 4 (December 1990): 744–65.

Moses, M., et al. "Environmental Equity and Pesticide Exposure." *Toxicology and Industrial Health* 9, no. 5 (1993): 913–59.

Muir, John. *Our National Parks.* Boston: Houghton Mifflin, 1901.

Musgrave, Richard A., and Peggy B. Musgrave. *Public Finance in Theory and Practice.* 3d ed. New York: McGraw-Hill, 1980.

Nader, Ralph. *Unsafe at Any Speed: The Designed-in Dangers of the American Automobile.* New York: Grossman Publishers, 1965.

Nash, R. "Part Two: The Progressive Conservation Crusade, 1901–1910." In *American Environmentalism: Readings in Conservation History*, pp. 69–71. 3d ed. New York: McGraw-Hill, 1990.

National Academy of Public Administration Report to Congress. *Setting Priorities, Getting Results: A New Direction for the Environmental Protection Agency.* Washington, D.C.: National Academy of Public Administration, 1995.

National Council of Churches. *Policy Statement on Racial Justice.* New York: National Council of Churches, November 1984.

Nieves, Leslie A. "Not in Whose Backyard? Minority Population Concentrations and Noxious Facilities Sites." Presentation for the American Academy of Science, Chicago, February 9, 1992.

———. "Regional Differences in the Potential Exposure of U.S. Minority Populations to Hazardous Facilities." Presentation for the Annual Meeting of the Regional Science Association, Chicago, November 19, 1992.

Okun, Arthur M. *Equality and Efficiency: The Big Tradeoff.* Washington, D.C.: Brookings Institution, 1975.

Pepper, David. *Modern Environmentalism: An Introduction.* London: Routledge, 1996.

———. *The Roots of Modern Environmentalism, The Croom Helm Natural Environment: Problems and Management Series.* London: Croom Helm, 1984.

Petulla, J. M. *American Environmental History.* 2d ed. Columbus: Merrill, 1988.

Portney, Kent E. *Controversial Issues in Environmental Policy: Science vs. Economics vs. Politics.* New York: Sage Publications, 1992.

Rhodes, E. L. "An Exploratory Analysis of Variations in Performance among U.S. National Parks." In *Measuring Efficiency: An Assessment of Data Envelopment Analysis*, ed. R. H. Silkman, pp. 47–71. New Directions for Program Evaluation, no. 32. San Francisco: Jossey Bass, 1986.

Rhodes, E. L., and L. Southwick. "Variations in Public and Private University Efficiency." *Applications of Management Science* 7 (1993): 145–70.

Roper Organization. *Roper Report 83-8.* Fieldwork August 1983, report October 1983.

———. *Roper Report 89-4.* Fieldwork March 1989, report June 1989.

——. *Roper Report 90-10.* Fieldwork October 1990, report January 1991.

——. *Roper Report 91-9.* Fieldwork September 1991, report January 1992.

——. *Roper Report 92-4.* Fieldwork March 1992, report July 1992.

——. *Roper Report 92-10.* Fieldwork October 1992, report March 1993.

——. *Roper Report 93-2.* Fieldwork January 1993, report April 1993.

Rosen, G. *A History of Public Health.* Baltimore: John Hopkins University Press, 1993; expanded ed. of *A History of Public Health.* New York: MD Publications, 1958.

Savas, E. *Privatizing the Public Sector: How to Shrink Government.* Chatham, N.J.: Chatham House, 1982.

Schroeder, Christopher. "The Evolution of Federal Regulation of Toxic Substances." In *Government and Environmental Politics: Essays on Historical Developments since World War II,* ed. Michael J. Lacey. London, Md.: Wilson Center Press, 1991.

Schultze, Charles L. *The Public Use of Private Interest.* Washington, D.C.: Brookings Institution, 1977.

Sherman, H. D. "Hospital Efficiency and Evaluation: Empirical Test of a New Technique." *Medical Care* 22, no. 10 (October 1984): 922–38.

Simon, Herbert. "Rationality as Process and as Process of Thought." *Proceedings of the American Economic Association* 68 (May 1978): 1–16.

Sinclair, Upton. *The Jungle.* 1906. Reprint; Urbana: University of Illinois Press, 1988.

Snow, Donald. *Inside the Environmental Movement: Meeting the Leadership Challenge.* Washington, D.C.: Conservation Fund, Island Press, 1992.

Solo, Kirkpatrick. *The Green Revolution: The American Environmental Movement 1962–1992.* New York: Hill and Wang, 1993.

Stanley, Harold, and R. G. Niemi. *Vital Statistics on American Politics.* 5th ed. Washington, D.C.: CQ Press, 1995.

Starling, Grover. *The Changing Environment of Business.* 3d ed. Boston: PWS-Kent Publishing, 1988.

Stokey, Edith, and Richard Zeckhauser. *A Primer for Policy Analysis.* New York: Norton, 1978.

Suksi, M. *Bringing in the People: A Comparison of Constitutional Forms and Practices of the Referendum.* Dordrecht: Martinus Nijhoff, 1993.

Tiemann, Mary. "Waste Trade and the Basel Convention: Background and Update." Report for Congress, Congressional Research Service, December 1998.

Tobin, Gary. "Suburbanization and the Development of Motor Transportation: Transportation Technology and the Suburbanization Process." In *The Changing Face of the Suburbs,* ed. B. Schwartz, pp. 95–111. Chicago: University of Chicago Press, 1976.

Trenberth, Kevin, ed. *Climate System Modeling.* Cambridge: Cambridge University Press, 1992.

Truman, David. *The Governmental Process.* New York: Knopf, 1951.

Tucker, W. *Progress and Privilege: America in the Age of Environmentalism.* Garden City, N.Y.: Doubleday, 1982.

Ullman, John E., ed. *The Suburban Economic Network: Economic Activity Resource Use and the Great Sprawl.* New York: Praeger Publishers, 1977.

United Church of Christ Commission for Racial Justice and Public Data Access. *Toxic Wastes and Race in the United States: A National Report on the Racial and Socio-Economic Characteristics of Communities with Hazardous Waste Sites.* New York: United Church of Christ Commission for Racial Justice, 1987.

United Nations. *Basel Convention on the Control of Transboundary Movement of Hazardous Wastes and Their Disposal (1989).* Washington, D.C.: Government Printing Office, 1991.

U.S. Department of Commerce, Bureau of the Census. *1990 Census of Population: Social and Economic Characteristics.* Washington, D.C.: Government Printing Office, 1993.

——. *Characteristics 1980.* PCo-1-B25. Washington, D.C.: Government Printing Office, August 1992.

——. *Current Population Reports,* "Educational Attainment in the United States . . ." Series P20, no. 476, pp. 96–98. Washington D.C.: Government Printing Office, 1994.

——. *Historical Statistics of the United States: Colonial Times to 1970, Part 1.* Washington, D.C.: Government Printing Office, 1975.

——. *U.S. Census of Population: General Population Characteristics 1990.* CP-1-26. Washington, D.C.: Government Printing Office, May 1992.

——. *U.S. Census of Population: General Population Characteristics 1980.* PCo-1-B25. Washington, D.C.: Government Printing Office, August 1992.

——. *U.S. Census of Population: Social and Economic Characteristics 1990.* CP-2-26. Washington, D.C.: Government Printing Office, September 1993.

U.S. Congress, Office of Technology Assessment. *The State of the States on Brownfields: Programs for Cleanup and Reuse of Contaminated Sites.* Washington, D.C.: U.S. Office of Technology Assessment, June 1995.

U.S. Environmental Protection Agency (EPA). *Brownfields Action Agenda.* Washington D.C.: U.S. EPA, January 25, 1995.

——. *Brownfields Title VI Case Studies: Summary Report.* Washington, D.C.: U.S. EPA, June 1999.

——. *Environmental Justice Report.* Washington, D.C.: Office of Environmental Equity, 1994.

——. *EPA Brownfields Assessment Demonstration Pilots.* Washington, D.C.: U.S. EPA, October 1996.

——. *Final Guidance for Incorporating Environmental Justice Concerns in EPA's NEPA Compliance Analysis.* Washington, D.C.: U.S. EPA Office of Federal Activities, April 1998, section 1.1.1.

——. *Headquarters Cultural Diversity Survey: Final Report.* Washington, D.C.: U.S. EPA, May 1993.

——. *Interim Guidance for Investigating Title VI Administrative Complaints Challenging Permits.* Washington, D.C.: U.S. EPA, March 1997.

——. Office of Environmental Justice Report. 1994.

——. *Summary of the Meeting of the National Environmental Justice Advisory*

Council: Washington, D.C., July 25–26, 1995. Washington, D.C.: U.S. EPA, 1995.

———. *U.S. Environmental Protection Agency Oral History Interview—1: William D. Ruckelshaus.* Washington, D.C.: U.S. EPA, January 1993.

———. *U.S. Environmental Protection Agency Oral History Interview—2: Russell E. Train.* Washington, D.C.: U.S. EPA, July 1993.

———. Office of Pollution Prevention and Toxics. *Toxic Inventory and Emissions Reduction 1987–1990 in the Lower Mississippi Industrial Corridor Report.* Washington, D.C.: U.S. EPA, May 14, 1993.

U.S. Food and Drug Administration. *The Story of the Laws behind the Labels,* part 1. 1906 Food and Drug Act, U.S. Food and Drug Administration, June 1981. Available at: http://vm.cfsan.fda.gov/~/rd/history.html.

U.S. General Accounting Office. *Siting of Hazardous Waste Landfills and Their Correlation with Racial and Economic Status of Surrounding Communities.* Washington, D.C.: General Accounting Office, 1983.

U.S. Office of Personnel Management. *Federal Civilian Workforce Statistics: Demographic Profile of the Federal Workforce.* Washington, D.C.: U.S. Office of Personnel Management, September 30, 1992.

"Views from the Former Administrators." *EPA Journal* 12, no. 9 (November 1985): 12.

Viscusi, K. "Reforming OSHA Regulation of Workplace Risks." In *Regulatory Reform: What Actually Happened,* ed. L. Weiss and M. Klass, pp. 234–68. Boston: Little, Brown, 1986.

Viv, A. "Toxic Trade with Africa." *Environmental Science Technology Journal* 23 (1989).

Voss, Jaap. "The Role of Local Planners and Decision-Makers in the Occurrence of Environmental Injustice." Ph.D. dissertation, Florida Atlantic University, 1997.

Wattenbert, B., with R. Scammon. "The Suburban Boom." In *North American Suburbs: Politics, Diversity and Change,* ed. John Kramer, pp. 71–81. Berkeley: Glendessary Press, 1972.

Weimer, David, and Aidan R. Vining. *Policy Analysis.* 2d ed. Englewood Cliffs, N.J.: Prentice-Hall, 1992.

———. *Policy Analysis.* 3d ed. Englewood Cliffs, N.J.: Prentice-Hall, 1999.

Weiss, L., and M. Klass, eds. *Regulatory Reform: What Actually Happened.* Boston: Little, Brown, 1986.

West, P., et al. "Minority Anglers and Toxic Fish Consumption: Evidence from a State-Wide Survey of Michigan." In *Race and the Incidence of Environmental Hazards: A Time for Discourse,* ed. B. Bryant and Paul Mohai, pp. 108–22. Boulder, Colo.: Westview Press, 1992.

Wildousky, Aaron. "No Risk Is the Highest Risk of All." *American Scientist* 67, no. 1 (January–February 1979): 32–37.

Wilson, William J. *The Truly Disadvantaged: The Inner City, the Underclass, and Public Policy.* Chicago: University of Chicago Press, 1987.

Woodhouse, Edward J. *The Policy-Making Process.* 3d ed. Englewood Cliffs, N.J.: Prentice-Hall, 1993.

World Bank. *The World Bank Atlas: Twenty-fifth Anniversary Edition.* Washington, D.C.: International Bank for Reconstruction and Development/World Bank, 1992.

Yount, K., and P. Meyer. "Bankers, Developers, and New Investment in Brownfields Sites: Environmental Concerns and the Social Psychology of Risk." *Economic Development Quarterly* 8, no. 4: 338–44.

Index

Acid rain, 92
Adeola, Francis O., 79
Affirmative action programs, 94
African Americans, 5, 27; cancer risks and, 24–25; dietary risks and, 28; environmental attitudes of, 75–80, 84; EPA and, 94–97; lead poisoning and, 28; in mainstream environmental movement, 72–73; in Noxubee County, 162; proximity to hazardous waste sites and, 25, 79–80, 122
African Americans for Environmental Justice, 176
Agenda setting, 11, 45–47, 59–60, 63–68, 187–195
Agriculture Department, 31, 73, 88–91, 128
Anderton, Douglas, 141–142
Animal rights activists, 90
ARCINFO, 241n6
Area-specific problems. See Geographic location specific problems
Asian Americans, 27, 28, 97, 123
Atgeld Gardens, 25
Atomic Energy Commission, 25, 91
Attitudes: within EPA, 87–90, 92–96; within mainstream environmental movement, 37, 52, 82; of minorities, 3, 7, 31, 34, 35, 75–80, 92–96
Audubon Society. See National Audubon Society
Authority, governmental, 52

Balanced Budget Act, 68
Bankhead, John, 176
Basel Convention on the Control of Transboundary Movement of Hazardous Wastes and Their Disposal, 243n17
Beck, Ulrich, 11
Been, Vicki, 141–142
Black Student Union (Indiana University), 178
Blackwell, Martha, 173
Bloomington (Ind.) City Council, 179
Bookchin, Murray, 34
Bowen, William, 141
Brooksville, Miss., 162, 167, 172
Brown v. Board of Education, 69
Browner, Carol, 181
Brownfields programs, 114, 198–199
Build absolutely nothing anywhere near anything policies (BANANAs), 9
Bullard, Robert, 41, 59, 82, 141–142, 200, 228n54, 235n2
Bureau of. *See under keywords (e.g., Solid Waste Management Bureau)*

California, 28
Cancer risks, 24–25
Caron, Judi A., 78–79
Census tracts/block groups, as assessment units, 20, 123, 129, 235n2
Charnes, Abe, 142
Chase, Steve, 34

257

Chavis, Benjamin, 59
Chemicals, toxic. *See* Dietary risks; Pesticide
    exposure
Chicago, Ill., 25, 200
Civil Rights Act, 24, 68–69, 88, 128, 181,
    198–199, 229n64
Civil rights movement, 47, 52–53, 56, 61–63,
    68–70, 116
Clean Air and Water Acts, 91, 195
Cobb, Roger, 63–64
Command-and-control regulatory approach,
    21, 60, 106, 113, 160, 191
Commerce Department, 100
Commoner, Barry, 50
Communities: classifications for assessment
    of, 20, 130–131; decision-making participa-
    tion and, 9, 18, 125, 189, 191–192, 194,
    197, 199–203; information access and,
    203–204
Community Right to Know Act, 204
Comprehensive Environmental Re-
    sponse, Compensation, and Liability
    Act (CERCLA), 170, 195
Computer simulations, 135
Conrad, Martin and Opal, 163–164
Conservation Foundation, 39
Conservation Fund, 73–74, 82
Conservation movement, 32, 36–40, 47
Consumer choice theory, 93
Consumer product safety, 57
Cooper, William, 142
Correspondence principle, 201
Counties, as assessment units, 20, 26–27,
    123, 129, 141, 235n2
Cove, Ariz., 25
Crawford, William A., 179

Data envelopment analyses (DEA), 135–136,
    142–145, 151–154; hypothetical illustra-
    tion of, 145–151; mathematical explanation
    of, 154–157
Deep ecology movement, 32, 34, 221n7
Deforestation. *See* Forests and deforestation
Delphi techniques, 132
Department of. *See under keywords (e.g.,
    Energy Department)*
Descriptive statistics, 134
Dietary risks, 28, 61, 99, 152–153. *See also*
    Fish toxicity

Downs, Anthony, 49–50
DRE Technologies, 175
Dunlap, Riley, 78–79

Earth First, 221n11
Economically specific problems, 7–8, 14,
    22, 124
Educational attainment, 52–53, 56, 126, 162
Efficiency, 16, 65–66, 107–108, 117, 142, 155
Ehrlich, Thomas, 164, 178–179
Elder, Charles, 63–64
Electric power generation, 50
Elite theory, 44, 228n45
Emelle, Ala., 162, 172
Employee recruitment, 82–83, 85, 93–95
Energy Department, 3, 31, 73, 89–90, 197
ENSCO of Arkansas, 175
Environmental Congress of Arkansas, 173
Environmental Justice Office (OEJ), 18,
    88, 99
Environmental movement, modern/main-
    stream, 1–2, 11–12, 30–42; agenda of, 30–
    31, 32–36, 41, 72, 74–75, 82, 188, 206;
    antiurban bias of, 37, 52, 82; as informa-
    tion provider, 39; minorities in, 6, 31, 33–
    34, 39, 41, 53, 72, 81–85, 207; new rights
    and, 116; origins of, 32–33, 36–38, 47, 50,
    90; problem recognition/classification and,
    60; recruitment for, 82; structural changes
    and, 52, 54
Environmental Protection Agency (EPA),
    3, 31; agenda of, 88, 91–92, 98–99, 190–
    191, 206; attitudes within, 87–90, 92–
    96; community participation and, 197,
    204, 242n6; definition development by,
    17–19; environmental justice research
    by, 5, 20–21, 89, 120–121; Executive
    Order 12898 and, 196–97, 229n64; as
    information provider, 99; minorities in,
    6, 31, 73, 96–97, 99–100; Noxubee haz-
    ardous waste disposal site and, 169–170,
    181; organizational structure of, 88, 91–
    92, 98–99; recruitment for, 89, 93–95;
    regional/field offices of, 95, 168; regula-
    tory responsibilities of, 113, 128; Title VI
    and, 198–199
Environmental Quality Department (Miss.),
    166, 173, 179
Equal Rights Amendment, 69

Equity/inequity, environmental, 1, 13, 16–
17, 107–108, 138, 193
Evolutionary policy processes, 44
Executive Order 12898 (Feb. 11, 1994), 70,
100, 195–197, 223n1, 231n2
Externalities, 109–111, 234n8
Eyestone, Robert, 223n3

Fair Housing Act, 128
Farm workers, 6, 14, 28–29, 61, 112–113,
123–124, 193, 236n1
Fauntroy, Walter, 58
Federal agencies: agendas of, 100, 206; envi-
ronmental justice research by, 20–21; as
information providers, 194, 203–204; insti-
tutional opposition in, 70; minorities in, 6,
73, 96–97; recruitment for, 82–83, 85. See
also individual agencies
Federal Insecticide, Fungicide, and Rodenti-
cide Act, 195
Federalism, 115–116, 189–190
Federated Technologies, Inc. (FTI), 164–
168, 170–173. See also Hughes Environ-
mental Systems, Inc.
Fish toxicity, 14, 29, 108, 123, 128. See also
Dietary risks
Food and Drug Act (1906), 58
Food and Drug Administration, 91
Food processing industry, 57–58
Fordice, Kirkwood, 181
Foreman, Dave, 34
Forests and deforestation, 33, 36, 92
Free rider problem, 109, 114–115
Friends of the Earth, 39, 40

Gaming exercises, 132
General Accounting Office (GAO), 58, 89, 99
Geographic assessment units, 20, 123, 129–
130, 141, 151–152, 235n2
Geographic location specific problems, 14,
22, 24–27, 102, 122–123, 220n29
Geographical information systems, 131, 134,
241n6
Global warming, 64–65
Goldman, Benjamin, 26–27
Gramm-Rudman-Hollings Budget Reform
Act, 68
Grassroots movements, 40–42, 82, 188, 207,
230n7. See also Organizations, formation of

Graves, James E., 181
Greenpeace, 39, 40, 173, 207
Group theory, 44

Haiti, 199
Hazardous waste sites, 5, 9–10, 108–109;
case study of, 161–185; free rider problem
and, 114–115; identification process and,
91; industrial location patterns and, 7–8,
23–24, 38, 55, 103; information costs and,
111–112; international dimensions of, 199;
mainstream environmental movement and,
33; measurements and, 138; minorities'
proximity to, 6, 23, 25–27, 54–55, 58–59,
79–80, 122, 141–142, 193; not in my backyard policies and, 60–61; transaction costs
and, 110
Health, Education, and Welfare Department, 91
Health care policies, 11, 68
Herald-Times (Bloomington, Ind.), 174
Hersey, M. R., 75
Hill, P. B., 75
Hird, John, 141
Hispanics, 27; EPA and, 94, 97; in main-
stream environmental movement, 72–73;
pesticide exposure and, 6, 14, 28–29, 61,
112–113, 123–124, 236n1
Home ownership, as assessment unit, 126
Horizontal equity, 218n3
Housing and Urban Development Depart-
ment, 197
Housing policies, 11, 25, 61
Houston, Tex., 59, 141–142, 200
Hughes Aircraft Corporation, 168
Hughes Environmental Systems, Inc.
(HESI), 168; partnership with Feder-
ated Technologies, Inc. (HESI-FTI),
168–169, 174–181

Incinerators, 171, 181–182
Income: distribution of, 50–53, 55–56; mea-
surement of, 126, 130, 191; racism and, 8,
11–12, 193
Incrementalism, 44
Indian reservations. See Native Americans
Indiana Daily Student, 174
Indiana University (IU), 163, 173–176
Indiana University Faculty Council, 178

Indiana University Foundation, 163–165, 173–175, 177–181

Indiana University Student Association (IUSA), 174–178, 182–183

Industrial location patterns, 7–8, 23–24, 38, 55, 103

Inequity. *See* Equity/inequity, environmental

Information: access to, 7, 55–56, 125, 191, 203–204; costs of, 109, 111–113; providers of, 39, 194, 203–204

Interior Department, 3, 31, 73, 88–91, 96, 128, 197

International environmental issues, 9, 10, 199, 209–210

Internships, 90

Issue attention cycles, 49–50

Jones, Charles, 45

Jung, C., 235n19

Kimberling, John F., 180

Kingdon, John, 224n8

Krause, Daniel, 77

Landfills. *See* Hazardous waste sites

Latinos. *See* Hispanics

Lead poisoning: EPA and, 99; exposure to, 6, 8, 14, 28, 61, 123; mainstream environmental movement and, 33; structural change and, 55

Left-right political/social alignments, 33–34

Legal Defense Fund (NAACP), 83–84

Legal Defense Fund (Sierra Club), 181

Legislative actions, 11, 68–70, 127–128, 190–191

Liability assessments, 112, 113, 114. *See also* Responsibility assessments

Linear programming computations, 136, 145

Local governments, 60, 115–116, 189–190, 199–203, 207

Locally undesirable land uses (LULUs), 9

Location specific problems. *See* Geographic location specific problems

Lower Bule reservation, 26

Mabus, Ray, 170

Macon, Miss., 162, 167, 172

Mahry, Sam, 176

Market efficiency, 65–66, 107–108, 117, 155

Market failure, 101–102, 108–116, 193

McWherter, Ned, 165

Measurement, 19–22, 102, 113, 118–136; causality and, 154; computer simulations and, 135; data envelopment analyses and, 135–136, 142–157; data reliability and, 153; demographic factors and, 126–127, 130–131; descriptive statistics and, 134; dimensions of, 125–129; economically specific problems and, 124; exposure factors and, 127; geographic units of, 20, 129–130, 141, 151–152, 235n2; of incomes, 126, 130, 191; multiple regression analyses and, 134–135; operations research and, 135–136; performance assessments and, 21–22; policy factors and, 127–128; population specific problems and, 130–131; problem type identification and, 24, 122–124, 132; qualitative analyses and, 132; responsibility/liability assessments and, 22–24, 191, 194–195; single risk assessments and, 21–22, 138–139, 142, 154; solution factors and, 128–129; univariate/multivariate statistical analyses, 134, 138, 152. *See also* Problem recognition and classification; Research studies, construction of

Medicare, 209

Merrill, Thomas, 180

Mexicans. *See* Hispanics

Michigan, 28, 29, 114

Milbraith, L., 75

Mineral rights, 175

Mississippi state government, 166, 169–170, 173, 175

Mississippi State University, 168

Mixed scanning policy processes, 44

Mohai, Paul, 78

Motaualli, Jim, 25

Muir, John, 36

Multiple regression analyses, 134–135

Multivariate statistical analyses, 134, 138, 152

Nader, Ralph, 57, 58

National Air Pollution Control Administration, 91

National Association for the Advancement of Colored People (NAACP), 63, 83–84, 222n15; Mississippi chapters of, 172, 177, 184–185

National Audubon Society, 40, 81, 88

National Council of Churches, 15

National Environmental Justice Advisory Council, 88

National Environmental Policy Act, 217n1

National Environmental Protection Act, 195, 197

National Park Service, 100

National Priority List (NPL). *See* Superfund sites

Native Americans: dietary risks and, 28; employment of, 96–97; EPA and, 94; hazardous waste sites and, 26, 29; land clearing by, 37; nuclear waste storage facilities and, 5, 7, 26, 124, 132, 236n1; uranium mining and, 25, 29

Navajos, 25, 29

Navy Department, 197

Neighborhood councils, 201–203

Netherland, Edward, 164–165

New York, 202

Nieves, Leslie A., 27

Nonappropriability, 110

Not in my backyard policies (NIMBYs), 7, 9, 60, 192

Noxubee County, Miss., case study: court rulings and, 181; EPA involvement in, 169–170, 181; major players in, 159, 162–166; proposal by FTI and, 164–170; statutory requirements and, 169–170; support and opposition in, 171–185

Noxubee County Committee, 174–175, 178–179

Nuclear waste storage facilities, 5, 7, 26, 79–80, 124, 132, 236n1

Numerics, 134

Occupational Safety and Health Administration (OSHA), 113, 160

Office of. *See under keywords (e.g., Environmental Justice Office)*

Offsetting compensations, 66–67, 152, 189, 192, 194, 226n34

Operations research analyses, 135–136

Organizations, formation of, 56, 61–63, 68. *See also* Grassroots movements

Pareto efficiency, 107, 155

Paternalism, 9, 177, 189, 191

Pennsylvania, 114

People of Color Environmental Leadership Summit, 188, 213–215

Pepper, David, 37, 226n29

Performance assessments, 21–22

Performance efficiency, 142

Pesticide exposure, 50, 99; analysis methods for, 153; farm workers and, 6, 14, 28–29, 61, 112–113, 123–124, 193, 236n1; free rider problem and, 114; regulation of, 91

Petrochemicals industry, 24–25

Pinchot, Gifford, 36

Political systems theory, 44

Polychlorinated biphenyls (PCBs), 58, 139, 175

Population density, as assessment unit, 126

Population shifts, 7–8; in Noxubee County, 162; from rural to urban, 57–58; from suburban to urban, 223n25; from urban to suburban, 37–39, 50–51, 52–53, 55, 58

Population specific problems, 14, 27–29, 102, 123–124, 130–131

Portney, Kent E., 221n2

Preservation movement. *See* Conservation movement

Privatization, 67

Problem recognition and classification, 1, 13–14, 22–29, 121–125; civil rights movement and, 56; of economically specific problems, 7–8, 14, 22, 124; of geographic location specific problems, 14, 22, 24–27, 102, 122–123, 220n29; of population specific problems, 14, 27–29, 102, 123–124, 130–131; public policy and, 6–9, 45, 57–61; time characteristics and, 22, 24–27, 124–125, 194. *See also* Measurement

Program Planning and Budgeting System (PPBS), 70

Property rights, 110, 113, 116–117

Protect the Environment of Noxubee County (PEON), 171–173, 176, 178

Public health movement, 38–40

Public policy analysis, 103–106, 186–187, 205

Public policy processes, 43–71; agenda setting and, 11, 45–47, 59–60, 63–68, 187–195; formalization and, 45–46, 68–69; implementation and, 45, 69–70; institutional opposition and, 70; organization formation, 56, 61–63, 68; paradigm shifts and, 10–12, 206–211; problem recognition/classification and, 1, 6–9, 13–14, 22–29, 45, 56–61, 121–125; program evaluation and, 45, 70–71, 106; stages of, 2, 44–45,

47–71, 187; structural changes and, 46, 49–57; theories of, 44

Purdue University, 163–164

Qualitative analyses, 132

Quality of life issues, 9, 50, 55–56, 60, 116–117

Racial composition, as assessment units, 126, 130

Racial Justice Working Group (National Council of Churches), 15

Racism, environmental, 6–7, 12, 89; charged in Noxubee County, 171, 175, 177, 184–185; definition of, 1, 13–16, 19

Radiation exposure, 25, 29, 91

Radiological Health Bureau, 91

Radon exposure, 124

Rational-comprehensive policy processes, 44

Recruiting environmental employees. See Employee recruitment

Recycling Sciences International, 166

Red Valley, Ariz., 25

Referendums, 201–203

Regulatory controls, 11, 68–70, 159–160; command-and-control approach and, 21, 60, 106, 113, 160, 191; costs and, 111, 113; Executive Order 12898 and, 70, 100, 195–197, 223n1, 231n2; federalism and, 115–116; flexibility/rigidity of, 21–22, 192, 195–196; violations/evasion of, 5, 26, 127–128

Research studies, construction of, 5, 8, 29, 64, 75, 93, 119, 140–141. See also Measurement

Residential patterns. See Population shifts

Resource Conservation and Recovery Act (RCRA), 169, 195

Responsibility assessments, 22–24, 60, 191, 194–195. See also Liability assessments

Rhodes, Edwardo Lao, 142

Rights, environmental protection, 1, 116–117

Romanticism, 37–38, 47

Roper Organization surveys, 75–77, 79–80, 84

Ruckelshaus, William D., 91

School funding, 10–11

Shultze, Charles L., 109

Shuqualak, Miss., 162, 167, 171, 172

Sierra Club, 39–40, 81, 88, 178, 181, 206, 221n9

Simic, Curtis, 164, 175

Sinclair, Upton, 57

Single risk assessments, 21–22, 138–139, 142, 154

Site specific problems. See Geographic location specific problems

Snow, Donald, 31

Social class, 11–12, 18, 66, 193

Solid Waste Management Bureau, 91

South Dakota Disposal System, 26

Space exploration program, 114

Special districts, 202–203

Standards. See Regulatory controls

Starling, Grover, 225n17

State Department, 100

State governments, 60, 100, 115–116, 169, 189–190, 207

Statistical analyses, 134–135, 152

Structural changes, 46, 49–57

Student Environmental Action Coalition, Indiana Chapter (SEAC), 173–175, 178

Subsistence fishing. See Fish toxicity

Summer intern programs, 90

Superfund Amendments and Reauthorization Action (SARA), 170, 195

Superfund sites, 5, 27, 125, 129

Surface mining operations, 128

Swine flu vaccines, 114

Tennessee, 165

Teton Lakotas, 26

Time characteristics, 22, 24–27, 124–125, 194

Time-series analyses, 135

Title VI, Civil Rights Act. See Civil Rights Act

Toxic Release Inventory (TRI), 91, 129, 138

Toxic Substances Control Act, 195

Toxic Wastes and Race in the United States, 14, 26, 141, 235n2

Train, Russell E., 91–92

Transaction costs, 65, 109–111

Transcendentalism, 37–38

Transportation Department, 197

U.S. Pollution Control, Inc. (USPCI), 166, 167

Uncertainty, 109, 113–114

Unemployment problems, 49
United Church of Christ (UCC), 14, 26, 59, 141, 235n2
United Nations, 243n17
Univariate statistical analyses, 134
Uranium mining, 25, 29
Urban League, 63, 222n15

Van Liere, K., 78–79
Vertical equity, 11, 218n3
Voting Rights Act, 68

Warren County, N.C., 58–59
Water Hygiene Bureau, 91
Water Quality Administration, 91
Wilderness Society, 39, 88
Women's rights movement, 69
Word, James, 165
Wright, B., 82
Wynn, Fred, 165

Zip codes, as assessment units, 20, 26, 141, 235n2

Edwardo Lao Rhodes is
Professor of Public and Environmental Affairs
at the Indiana University School of Public
and Environmental Affairs.